T0270889

FOOD PROCESSING AND PRESERVATION

This book provides an exhaustive coverage on all the types of food products-fruits, vegetables, cereals, dairy and meat. The initial two chapters provide a brief introduction to food processing and preservation and their importance in employment generation. The next chapter (chapter 3) is devoted to the processing and preservation of fruits and their products. Subsequently, the succeeding chapters deal with the processing and preservation methods for vegetables (chapter 4), cereals (chapter 5), dairy (chapter 6) and meat (chapter 7). This book is written and compiled in a simple and lucid language so that students can easily comprehend the information.

Dr. H. R. Naik is presently working as Dean Research, Islamic University of Science and Technology, Awantipora and Professor & Head, Department of Food Technology, Islamic University of Science and Technology, Awantipora, J&K. He has also served as the Head, Division of Food Science and Technology, SKUAST-Kashmir. He has been the Student's Welfare Officer and Placement Officer at SKUAST-Kashmir.

Dr. Tawheed Amin is presently working as Assistant Professor-cum-Junior Scientist (Food Science and Technology) in the Division of Food Science and Technology of Kashmir (SKUAST-Kashmir). He has been a Food Safety Officer in the Drug and Food Control Organization, Government of Jammu & Kashmir.

FOOD PROCESSING AND PRESERVATION

H.R. Naik

Dean Research,
Islamic University of Science and Technology,
Awantipora-192122 (J&K)

and

Tawheed Amin

Assistant Professor-cum-Junior Scientist,
Division of Food Science and Technology,
Sher e Kashmir University of Agricultural Sciences and Technology of Kashmir,
Shalimar-190025 (J&K)

CRC Press
Taylor & Francis Group
Boca Raton London New York

CRC Press is an imprint of the
Taylor & Francis Group, an **informa** business

NARENDRA PUBLISHING HOUSE
DELHI (INDIA)

First published 2022
by CRC Press
2 Park Square, Milton Park, Abingdon, Oxon, OX14 4RN

and by CRC Press
6000 Broken Sound Parkway NW, Suite 300, Boca Raton, FL 33487-2742

© 2022 Narendra Publishing House

CRC Press is an imprint of Informa UK Limited

The right of H.R. Naik and Tawheed Amin to be identified as authors of this work has been asserted in accordance with sections 77 and 78 of the Copyright, Designs and Patents Act 1988.

Print edition not for sale in South Asia (India, Sri Lanka, Nepal, Bangladesh, Pakistan or Bhutan).

British Library Cataloguing-in-Publication Data
A catalogue record for this book is available from the British Library

Library of Congress Cataloging-in-Publication Data
A catalog record has been requested

ISBN: 978-1-032-15247-9 (hbk)
ISBN: 978-1-003-24325-0 (ebk)

DOI: 10.1201/9781003243250

Contents

ISLAMIC UNIVERSITY OF SCIENCE & TECHNOLOGY AWANTIPORA, KASHMIR.

Prof. Mushtaq A. Siddiqi
Vice Chancellor

FOREWORD

Food preservation is an action or a method of maintaining foods at a desired level of properties or nature for their maximum benefits. In general, each step of handling, processing, storage, and distribution affects the characteristics of food, which may be desirable or undesirable. Thus, understanding the effects of each preservation method and handling procedure on foods is critical in food processing. The processing of food is no longer as simple or straightforward as in the past. It is now moving from an art to a highly interdisciplinary science. Food processing needs to use preservation techniques ranging

from simple to sophisticated; thus, any food process must acquire requisite knowledge about the methods, the technology, and the science of mode of action. Keeping this in mind, this edition has been developed to discuss the fundamental and practical aspects of most of the food preservation methods.

I am delighted to write the foreword for the book **"Processing and Preservation of Foods"**. Food processing and preservation is one of the inseparable parts of human life. This book provides a widely useful compilation of almost all small and medium scale processing and preservation methods for different food products. This book highlights the easy and simple techniques for processing and preservation of food products. This book is an important resource for teachers, students, industry professionals and house makers.

It is my hope and expectation that this book will be of the benefit to students, scientists, academicians, homemakers and professionals dealing in the area of food processing and preservation. Clearly written, well organized and enormously practical, it should be in every college and university library.

m.a Siddiqi

Prof. Mushtaq A. Siddiqi
Vice-Chancellor
IUST, Awantipora

1, University Avenue, Awantipora, Kashmir – 192122, J&K. **Contact:** 01933-247265   **Fax:** 01933-247248
Website: www.islamicuniversity.edu.in.   **Email:** vc@islamicuniversity.edu.in.

Preface

Food processing and preservation is an important discipline of food science and technology since it adds value to the food and maintains its quality. Therefore, one needs to gain a proper perspective and insight into the subject of food processing and preservation. Thus, this book was designed with an aim to provide some basic and simple processes and procedures to process and preserve the food.

This book provides an exhaustive coverage on all the types of food products-fruits, vegetables, cereals, dairy and meat. The initial two chapters provide a brief introduction to food processing and preservation and their importance in employment generation. The next chapter (chapter 3) is devoted to the processing and preservation of fruits and their products. Subsequently, the succeeding chapters deal with the processing and preservation methods for vegetables (chapter 4), cereals (chapter 5), dairy (chapter 6) and meat (chapter 7). This book is written and compiled in a simple and lucid language so that students can easily comprehend the information.

This book is an ideal book for diploma, undergraduate and post graduate students studying food science and technology. Besides, this book will provide the practical exposure to students.

We wish to express our sincere gratitude to our families for their understanding and patience while writing this manuscript. Finally, our heartfelt and sincere thanks go to the publishers of this book, Jaya Publishing House, Delhi for meticulously processing this manuscript.

H. R. Naik

Tawheed Amin

CHAPTER - 1

INTRODUCTION TO PRESERVATION OF FOODS

Food preservation consists of the application of science-based knowledge through a variety of available technologies and procedures, to prevent deterioration and spoilage of food products and extend their shelf-life, while assuring consumers a product free of pathogenic microorganisms. Shelf-life may be defined as the time it takes a product to decline to an unacceptable level. Deterioration of foods will result in loss of quality attributes, including flavor, texture, color, and other sensory properties. Nutritional quality is also affected during food deterioration. Physical, biological, microbiological, chemical, and biochemical factors may cause food deterioration. Preservation methods should be applied as early as possible in the food production pipeline and therefore, include appropriate postharvest handling before processing of both plant and animal foods (Figure 1).

Processing techniques usually rely on appropriate packaging methods and materials to assure continuity of preservation. Handling of processed foods during storage, transportation, retail, and by the consumer also influences the preservation of processed foods.

All of the processes used for preserving food at home are based on the principle of preservation.

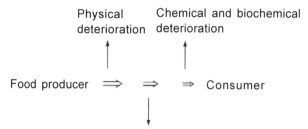

Figure 1. Losses in the food production pipeline

CHAPTER - 2

ROLE OF FOOD PROCESSING/ PRESERVATION IN INCOME AND EMPLOYMENT GENERATION

INTRODUCTION

The diversification and modernization of the present agricultural and other related activities supported by efficient on and off farm processing of the commodities for the purpose of value-addition is expected to increase food production and create employment and income generation. Adding value to food commodities after harvest is also aimed at minimizing the losses during storage and to maintain the quality of product. Efficient post-production practices, particularly the preservation and processing of agricultural and allied produces may bring a wide range of benefits to the people in this country, generating job opportunities by opening up village-level processing units. The goals of post-harvest and food processing technology are loss prevention as well as adding value to the harvested biomass, which result in more income to the farmers/processors and better quality produce provided to the consumers. Post-harvest and food processing technology are commodity- and location-specific and it is done at home, village and/or cottage levels at small and large industrial scale. On-farm post-harvest storage and primary processing integrated with production technology help to generate more employment opportunities and additional income for rural people. Minimization of the post-harvest losses is an important means to increase per capita food availability. It also helps to generate more employment and income. Investment in post-harvest measures is more economical and time saving than in productivity to obtain the same amount of a particular commodity. Furthermore, post-harvest measures automatically add value to the raw commodities as they pass on their marketing channels. Adoption of post-harvest technologies and value additive measures are very strong tools for rural and social development through employment and income generation. Development and adoption of efficient value addition practices will enhance national food supply and sustain food security even at the household level. Fruits and vegetables processing industries have a good deal of potential in serving the rural economy. First, it helps in generating more employment for rural people. It will also check mobility of rural masses towards urban areas in search of employment. Employment opportunities offered by agro-processing industries are plenty to the farm population and entrepreneur seeking self-employment. Cottage scale units particularly offer self-employment opportunities. Traditionally women handle food and are familiar with skills of food processing. In order to improve the status of living of woman and rural food processing, low cost appropriate fruit and vegetable processing technologies offer excellent opportunities for production of processed foods. The improvement of status of socially backward and landless labor classes will be possible only through providing non-farm employment at their doorsteps. This will generate a sense of security and confidence amongst rural people for overcoming uncertainty in agricultural income and providing self-employment to the land-less labor. The locally available untapped resources should be used effectively.

Food processing as a scientific and technological activity covers a broader area than food preparation and cooking. It involves the application of scientific principles to slow down the natural processes of fruit decay caused by microorganisms, enzymes in the food or environmental factors such as, heat, moisture and sunlight- and so, preserve the food. In developing countries like India, food processing is a method of generating employment and family incomes but producers have to compete with others in the same country and with imported products where in packaging and preservation play a great role. Processing of foods involves methods of business planning, work organization and quality assurance that are likely to be unfamiliar to traditional preservers but which are essential to ensure successful and profitable production.

Issues to be considered when addressing different objectives of food processing

A) Issues that affect food security and nutritional improvement programmes

- Nutrition education
- Health and hygiene training
- Improved communication
- Confidence building measures
- Credit support systems
- Improvements to equipments and tools
- Improved infrastructure and transport
- Seed banks and other sources of agricultural inputs.

B) Issues that affect enterprise development programmes

- Suppliers of specialist equipment Market awareness and consumer preferences.
- Marketing strategies, promotion and packaging.
- Methods of financial control.
- Developing trust with suppliers and retailers.

- Quality assurance.
- Hygiene and sanitation for production
- Food legislation.
- Taxation and business legislation.
- Training in management and business planning.
- Finance and credit suppliers.
- Staff training in production technologies.
- Ingredients.

Status and Scope of Food Business in India

The growth rate of food production being faster than population growth has led to development in agro industry particularly Food Industry in India through the successive five year plans. Processed foods have slowly won the hearts of consumers in the recent years and the production is increasing but still not sufficient to meet the demands of people. The food industry has to dedicate itself from raw material selection, mode of preparation, process to be adopted, packaging system to be used, the quality control of raw material, final product as well whole operations involved to meet the local and international standards.

Evolution of Food Processing Industry

The food processing industry in India has evolved from different phases, basic development phase-1947 to 1970, transformation phase from 1970-1990, early growth phase of 1999- 2002 and consolidation phase 2003 and beyond as influenced by economic growth, changing consumer attitudes and government policies. The potential for this industry in India is enormous provided there is good infrastructure, better cultivation practices, systematic processing approach with the backing of resources and policies. Today, India is the world's leading producer of foods starting from sugar, tea, milk to fruits and vegetables. India produces 601 million tons of food as against 608 million tonnes in US. The food industry structure reveals that only 55% of food production is contributed by small scale and organized sectors while 42% is being produced by the unorganized sector. Indian Food Industry is fifth largest, employs 19% work force, 14% industrial output, 5.5% GDP. The present worth of food business in India is Rs. 8,70,000 crores. The composition of Indian food industry reveals that major industry consists of oils and fats (36%), followed by dairy products (16.9%), cold beverages (15.6%) and beverages (15.7%). Industrial contribution is mainly for primary processed preparation or formulations.

PROCESSING INDUSTRY-RAW MATERIAL (MILLION TONES)

Raw Material	Production
Food grain	205 (2)
Cereals	185(2)
Pulses/Legumes	15
Oil seeds	25
Sugarcane	280 (1)
Horticultural Crops	148 (2)
Milk	88 (1)
Fisheries	5.5 (3)
Livestock (million)	500 (1)
Poultry birds (million)	1500 (5)
Egg (billions)	38 (5)

● Figures in parentheses indicate ranking of India in global production

Dimension of Industry

In India, processed food sector constitutes 38% of the Indian food business and has grown at the rate of 20% in the last 3 years. Processed food industry is 2.5 times faster than the agricultural sector yet the value addition to agro food commodities in India is only 7% as compared to 23% in China, 45% Philippines, 180% in USA and 188% UK. In India, economic losses due to poor postharvest infrastructure and lack of processing facilities are valued at Rs. 77, 000 crores. There are 30-40%

losses in fruits and vegetables, 10-15% in food grains, 5-7% in milk, 5% in meat and 10% in eggs.

Profile of the Industry

The processing of foods is an age old process starting with pounding of paddy to get rice, sun drying of fruits and vegetables, manual extrusion of gelatinized cereals, pickling of mango/citrus fruits, roasting of spices etc. The advancement of technology and engineering has transformed the conventional method by pounding of paddy to cleaning of paddy, grading, dehusking, polishing and utilization of byproduct of milling. Similarly, wheat milling through conditioning, breaking, pulverizing through reduction rolls instead of *chakki* grinding. Corn milling is done by using entoleter machine for degerming and dehulling process followed by break rolls and shifters. Millets are ground to whole flour besides brewing, malting and roasting. Sorghum and maize are converted into flours and flakes. The cereal processed products are on an increase and the production is to the tune of 30 lakh tones. There are around 820 roller flour mills and 35088 modern rice mills. The production of bread is 15.20 lakh tonnes and that of biscuits is 15.3 lakh tons with the growth rate of 8% per annum. Packaged "*atta*" marketing is on the increase.

Oat and wheat bran is used as a functional fiber. Rice bran production is 34 lakh tonnes and is well utilized for oil production. India is the second largest producer of fruits and vegetables. There is a steady growth of processed fruit and vegetable. During 2019, India's total annual fruit and vegetable production was 98 and 185 million metric tons respectively. Though range of processed products is quite large, yet manufacturing is confined only to few items *viz.* pickles, chutneys, murrabbas, fruit juices, squashes, jam etc. while dehydration has limited value addition in the form of basic infrastructure. Produce from individual farm passes through the hands of at least seven intermediaries before it reaches the retailers. Better storage and handling facilities at farm level and reduction in large number of intermediaries reduces the losses. This sector is growing at the rate of 15% per year. Though the volume of processed fruits and vegetable consumed in India is very low, yet the export potential is very large. India has the highest level of milk production and consumption amongst all the countries with annual production of 186 million tons. About 13% milk produced is processed in organized sector and about 46% of milk is consumed in the liquid form. As per FAIDA report, operation flood has linked 8.4 million farmers in 200 districts into 70,000 milk societies and these societies supply milk to 170 milk unions for processing and marketing. The dairy industry is growing at the rate of 5 percent per year due to co-operative movement. The production of oilseed is 25 million tones. The domestic requirement is satisfied by importing 1.50 to 2.00 million tons of edible oils. Newer oilseeds used are sunflower, soybean, cotton seed and palm. Still supplies are running short by 2.1% hence there is a need to increase area under oilseed cultivation.

India is the largest producer (25 lakh tones), consumer (22.8 lakh tones) and exporter (2.5 lakh tones) of spices in the world. The exports of spices are worth more than 1260 crores rupees. More than 50% of total meat production of 4.11 MT is from buffalo and cattle. Almost 55% of fish produced in India is from the marine production. Processing of marine products into canned and frozen forms is carried out for export. Poultry growth is on increase with increased availability of eggs and broilers.

Infrastructure Status

Many of the specific infrastructure needs of food industry are inadequate or non-existent. There is a lack of rapport between industry and agriculture, underdeveloped warehousing and cold storage facilities, inadequate transport infrastructure, perennial shortage of quality power, insufficient initiatives in design and fabrication and little equipment for processing and packaging and absence of Research and Development. All these have resulted in slow pace of growth.

Quality and Safety Aspects

In India, food industry is handicapped by many impediments in keeping up with the aspirations and expectations of consumer community. Our industry depends on raw material not specifically suited to obtain good quality product. There is a lack of in-house quality control facilities, reluctance to adopt ISO 9000 Management System and HACCP Safety norms, ineffectiveness of BIS standard systems and high cost and delay in safety and quality evaluation.

MAJOR BRAND OWNING INDUSTRY IN FOOD SECTOR

MULTINATIONAL	INDIAN
Hindustan Lever Ltd	Parle Biscuits
Nestle India	Nutrine Confectionery
Cadbury India	Godrej Foods Ltd.
Coca cola India	MTR Foods Ltd
Pepsi cola India	Haldiram Foods Ltd
Britannia Industries	Parry confectionery
Smith Kline Beecham	Norton confectionery
Cargill India	Amul (NDDB)
ITC Ltd	Gits Products
Kellogg's India	Vadilal Ice Cream
International Best Food	Dabur Foods

Food Laws and Regulations

In India, antiquated food laws are PFA, 1955. Amended from time to time it is handicapped in upgrading the quality in tune with international development in Food Technology. Even quality assessment facilities are underdeveloped. Enforcement agencies and inspection and quality monitoring system is inadequate and result in delays in clearing import and export shipments, delays in harmonizing the food standards and restrictions with those evolved by WHO-FAO Alimentarius Commission. Govt. of India has also planned to bring stricter food Laws under Food Safety Act.

The Packaging Scenario

Packaging material produced in India is 5.20 million tones which is only 0.40% of world production and consumption. About 43% of material is consumed by food processing industry. Innovation and development in packaging is taken care of Indian Institute of Packaging, Mumbai and CFTRI, Mysore.

Personnel Situation

Universities and Training Institutions are spread throughout the country to develop Food Technologist and Engineers with sound knowledge of Food Science and Technology. Chemists are trained for analysis. Training programmes for ISO-9000 Management Systems, HACCP and other are organized by institutions from time to time.

Future

- Consumer plays a significant role in deciding the development of food processing industry.
- Health awareness, understanding of food and nutrition has resulted in increased consumption of processed foods.
- Fast foods centers are on rise. The time saving, easy to prepare foods that fit in buying capacities is gaining significance.
- Organic Foods- world market 26 billion $. India Rs. 75 –100 crores.
- Investment needs for next 10 years Rs. 1, 50,000 crores for 35% value addition.
- Organized retailing in food 2% to grow at the rate of 20%/ year.

		Rs. (in crores)
Bakery	=	10,000
Branded Atta	=	15,000
Fresh poultry	=	27,000
Sugar	=	24,000
Packaged milk	=	36,000

- Packaged food market in the world is estimated to be of Rs. 680 billion $ and International Exports Rs.400 billion out of which Indian Exports of Rs.14.6 billion $ with Indian exports growth 8%.
- Market to grow by 200% in next 3-4 yrs.

Scope for Jammu and Kashmir

There is enough scope for minimization of post harvest losses (30- 35 %) of fruits in the UT of J&K by appropriate value-addition technologies which can ensure food security, income and employment generation to the society. The rich horticultural resources have great potential for the development of fruit processing industry in J&K state. The area under fruits in the state is estimated to be 1.2 lakh hectares and fruit production has touched the mark of 10.97 lakh metric tons. J&K state produces temperate fruits like apple, walnuts, pear, almonds, peach, plum, cherries and quince in the temperate belts while as mango, lime and banana are grown in the sub-tropical region of Jammu. In Uri and Ramban areas, olive plantation has come up which can serve as base for providing olives for processing industry. The cold arid region of Ladakh is well suited for fruits like apricots, apples and sea-buckthorn is also grown in natural form.

Areas of Income and Employment Generation Through Food Processing

- Fresh fruit handling, transportation and storage
- Pack house facilities for pre-cooling, grading, and packing and refrigerated transportation of cherries, apples, pear, plums and strawberries for export.
- Creation of cold storage facilities on community basis with assured electric supply.

Value Addition

Food	Value added product
Apple	Juice, concentrate, powder, jam, jelly, sauce, pickle, toffee, bar, Dehydrated rings, baby foods, cider, canned slices, preserve.
Pear	Canned pear, pear juice, dehydrated pear halves, pear nectar.
Cherry	Canned cherry, pulp, squash, candy, nectar, jam, bar, RTS beverage
Strawberry	Canned strawberry, pulp, nectar, squash.
Plum	Pulp, sauce, nectar, jam, canned plum.
Quince	Jam, preserve, pickle, jelly, *tuti- fruiti*.
Walnut	i. Value addition by de-hulling, bleaching, drying and packaging of Whole nuts.
	ii. Value addition by drying, grading and vacuum packaging of Whole nuts.
	iii. Walnut cake, snack foods.
Apricot	Pulp, concentrate, jam, jelly, toffee, bar, dehydrated fruits, baby foods, canned, apricot kernel syrup.
Mango	Pulp, squash, nectar, jam, bar, RTS beverage, pickle.
Lemon	Lemon Juice, squash, nectar, pickle.
OliveMushroom	Oil & pickle Canned mushroom, dehydrated mushroom & pickle
Honey	Value addition by purification/filtration and packing
Garlic and onion	Dehydrated products
Tomato	Tomato puree, sauce, ketchup and dried tomato powder
Cereals	• Rice milling, wheat flour milling, and packaging in consumer packs.

Food	Value added product
	• Bread, biscuits and local traditional bakery units. Extruded snacks.
Pulses	Pulse milling and value addition by appropriate packaging in consumer packs.
Oil seeds	Edible oil extraction from *sarson* and rice bran oil, value addition by refining and modern packaging.
Spices	Chillies, mint, coriander, methi, value addition by drying, grinding and packaging. Chilies value addition by making dried powder or by making oleoresin.

Value Added Centers Based on Packaging Industries

• There is enough scope for making other forms of packaging boxes other than wood in the state keeping in view the cost of wooden boxes and strict laws for tree cutting.

• Thus, wooden boxes can be replaced by card board and plastic based boxes to meet the requirements of fruit/food industry

CONCLUSION

India has comparative advantage in labor-intensive agricultural production and processing and could potentially be producing a wider variety of horticultural and agro-processing products. Post-harvest technology activities are likely to expand at a much faster rate than what can be stipulated at the present scenario with the dramatic increase in the production of commodity items as required by the future demand. Therefore, research and extension capabilities must be built in order to undertake dynamic endeavor to keep pace with the production. Successes in development of post-harvest processing technology and industries are hindered by a growing number of constraints. The main constraints are:

- Lack of readily available modern machinery, equipment, and technologies suited to local condition;
- Absence of reliable supply of raw materials;
- Poor managerial skills;
- Increased reliance on the part of many producers on imported raw materials (preservatives, color, flavor, emulsifier, etc.) which results in increased cost of production;
- Most of the modern and special processing equipment have to be imported which are expensive and difficult to maintain;
- For cottage and small-scale industries, promotional activities are limited due to high cost of publicity in mass media;
- Uncertainty of processed product market in the domestic market, there is stiff competition from multinational companies;
- Low and fluctuating nature of demand, high taxation and absence of transport make serious bottlenecks in marketing products;
- Lack of forward and backward linkage industries storage facilities; and
- Lack of standard packaging facilities.

RECOMMENDATIONS

Agriculture continues to be the mainstay of the economy of India. It remains as the major source of rural employment and the driving force behind its economic growth. The entry of this country into the World Trade Organization has opened up opportunities for new markets of its products and on the other hand exposed the country to greater competition. In the markets of the world, consumers demand products, which are perceived to be of higher quality than those grown in this country. There is, therefore, an urgent need to grow agricultural products of higher qualities for marketing, distribution and trade. The policy issues for a sustainable and reliable development of post-harvest technologies in the country to increase food production generate employment and income can be summarized as:

- Strong urban-rural linkage should be developed to ensure sustainable development of agro- industrial base in the villages;
- Development of alliances between large enterprises usually urban based and small and medium scale enterprises (SMEs) to be created at the rural level for creating dynamic agribusiness sector in rural areas;
- Establishment of SMEs should be at the forefront of the agribusiness sector, adding value to domestic raw materials, generating employment;
- Agricultural education has to be modified with incorporation of modern concepts and technologies. Selected faculty members from agricultural research and educational institutes should be trained in agro-business and curricula development;
- Upgrading of quality for better competition and marketing is needed;
- For successful implementation and management of a value addition enterprise women participants should be empowered and gender issue should be properly dealt;
- Extending existing policy of providing more financial assistance to processing industries;
- Providing financial and technical support for the development of packaging industries;
- Providing custom relief or tax rebates for importing specialized transport vehicles with cooling system for carrying fresh commodities to urban areas or to the processing industries; and
- Supporting specialized research programmes for the development of the suitable processing technologies as well as producing quality raw commodities.

CHAPTER - 3

PROCESSING AND PRESERVATION OF FRUITS AND THEIR PRODUCTS

INTRODUCTION

Nutritional wellbeing is a sustainable force for health and development of people and maximization of human genetic potential. From the beginning of human history, food has been considered as the major factor in maintaining wellbeing and health of individuals. Active ingredients in food, which are effective in promoting human health, include amino acids, fats, dietary fiber, antioxidants, pigments, vitamins and minerals which are present in different food groups such as pulses, cereals, legumes, oilseeds, fruits and vegetables.

Among all these food groups, fruits and vegetables play a significant role in human nutrition, especially as a source of vitamins, minerals and dietary fiber. The different fruits and vegetables like carrots, tomatoes potatoes, ginger, green leafy vegetables and the like are important protective foods, because of their nutritional value and antioxidant properties. Value addition of such fruits and vegetables by formulation of different value added products are an important source of nutritional security.

IMPORTANCE OF FRUITS AND VEGETABLES

Fruits and vegetables, as well as roots and tuber crops are among the best sources of calories, natural vitamins and minerals essential for healthful living. Green leafy vegetables such as amaranth, spinach fenugreek leaves, chenopodium album (bathua), mint etc., and roots and tubers such as carrots are rich sources of beta carotene, the most important precursor of vitamin A in human nutrition. Beta carotene has an important antioxidant fraction, which deactivates oxygen and free radicals, and thereby, gives protection against cancer. Vitamin A is essential for normal growth and vision, reproduction, maintenance of epithelial cells, immune properties, and its deficient intake results in a decreased blood levels and low levels in serum, showing sign of vitamin A deficiency. It has been observed that the current availability of fruits and vegetables meets only about half of the requirement of different vitamins and minerals and hence, there is a need to boost the production and handling of vegetables and fruits, to enhance the nutrition of rural and urban poor. Therefore, it becomes necessary that the processing of vegetables must be augmented by developing such techniques, which would be not only feasible but also would suffice to produce economic quality products. This makes availability of off season vegetables round the year. In India, less than 2 percent of the vegetables of the total production are being processed as against 70 percent in Brazil and 83 percent in Malaysia. The most common method for preservation of fruits and vegetables is the dehydration method. The vegetables can be dried by hot air-drying method for small scale operation or by conventional tray drier or vacuum drier, and at home level can be processed by sun drying method. These dehydrated forms of vegetables may be eaten as such or may be consumed in several forms, without affecting its nutrition and palatability. Vegetable powders such as carrot, tomato and fenugreek leaves powder can be prepared with simple technologies and can be incorporated in traditional food preparations, thereby adding value to the products and attaining food and nutrition security both.

ROLE OF PROCESSING

Vegetables are classified as green leafy vegetables, roots and tubers and others. Carrots among roots and tubers and fenugreek leaves among green leafy vegetables grown in winter season occupy an important place. These vegetables are rich sources of beta carotene and are generally marketed fresh and consumed as raw or cooked vegetables. Due to the seasonal availability, efforts are made to process the vegetables in large quantity to extend the shelf life and to make them available during rest of the year and in the areas, where they are not available. Preservation of vegetables by processing not only involves the inhibition of microbial growth, but also preserves their color, texture, flavor and nutritive value. The vegetables can be processed into different forms to extend their shelf life such as powders, grits, flakes, pulp, puree, etc.

VALUE ADDITION OF FRUITS AND VEGETABLES

Various research studies have been conducted on value addition of different fruits and vegetables such as carrots, tomatoes, potatoes, sweet potatoes, ginger, green leafy vegetables and spices, wherein different value added products have been developed from them, which are important sources of nutritional security. These are being described one by one as given below;

VALUE ADDITION OF TOMATOES

Tomatoes are one of the most widely used and versatile vegetable crops, ranking second in importance to potatoes in many countries. Tomatoes are important both for its large consumption and richness in health related food components. Tomato (*Solanum lycopersicon*) is an herbaceous plant of Solanaceae family, which is one of the most popular protective foods, because of its lycopene content, outstanding nutritive value, antioxidant properties and a powerhouse of medicinal properties. It is a rich source of minerals like calcium, magnesium, phosphorous, iron, sodium, potassium and vitamins especially A and C. Tomatoes are consumed mainly as a raw staple food, as an ingredient in different types of food products and in the form of processed products such as powder, tomato juice, paste, puree, sauce, etc. This horticultural crop is an excellent source of health promoting compounds, being a balanced mixture of minerals and antioxidant vitamins including vitamin C and E, as well as rich in lycopene, beta carotene, thiamine, riboflavin, niacin, lutein and flavonoids such as quercetin. The main antioxidants in the tomatoes are the carotenoids specially lycopene which have the highest lycopene levels among fruits and vegetables, ascorbic acid and phenolic compounds. In addition to lycopene, violaxanthin, neoxanthin, lutein, zeaxanthin α-cryptoxanthin, β-cryptoxanthin, carotene, neurosporene, phytoene and 5, 6-epoxides are other carotenoids commonly cited in tomatoes and tomato derived products. Among the different carotenoids, lycopene, is the most abundant in human serum, with important antioxidant activity involved in prevention of several types of cancer and degenerative diseases such as cardiovascular diseases. The production of tomato, an important horticultural crop of India has increased enormously during past few decades, which emphasize more on processing and preservation of tomatoes, thereby ensuring better availability and utilization during off season. India is the fourth largest producer of tomatoes, accounting for 6.6 percent of the world production and second largest in acreage. However, due to lack of proper processing, storage and transportation facilities, enormous quantities of tomatoes are lost during the peak harvesting season in India. Being a perishable crop, tomatoes cannot be stored for a longer time, hence proper processing and storage in some preserved form during seasons of glut will ensure its availability and utilization during deficiency period. Hence, processing of tomatoes in different forms as preferred by the consumers, with long shelf life involves low cost of production. Processing of fresh tomatoes can be done to prepare the following value added products such as;

- Tomato pulp
- Tomato puree
- Tomato paste
- Tomato flakes
- Canned tomatoes
- Tomato ketchup
- Tomato soup and sauce
- Tomato powder
- Dehydrated tomato

Therefore, replacement of fresh tomatoes for example, with tomato powder can facilitate the processing sector with daily cuisines and preparation during off season. Tomato powder can be used in processed products, such as soup mixes and confectionary items.

TOMATO POWDER

Select fresh tomato

Wash and clean

Cutting slices
(4 mm, 6 mm and 8 mm thickness)

Pretreatment with KMS, Sodium Benzoate
and control (untreated)

Drying slices in cabinet tray dryer (65°C)

Tomato flakes cool at room temperature

Flakes crush in mixer grinder

Tomato powder package
(aluminum foil and LDPE)

Storage

TOMATO COCKTAIL

Recipe:

Tomato juice	:	4.5 kg
Cloves (Headless whole)	:	1.5 g
Cumin	:	1 g
Black pepper	:	1 g
Cardamom	:	1 g
Red chilli, finely ground (Kashmiri)	:	0.25-1 g
Cinnamon (Broken)	:	0.25 g
Vinegar (5% acetic acid)	:	203 g
Common salt	:	45 g

Method:

1. Select fully ripe and healthy tomatoes.
2. Wash them thoroughly in fresh water.
3. Remove and discard the green and blemished portions.
4. Cut the fruit into small pieces and cook until soft. Crush with a wooden ladle while cooking.
5. Rub the mass through a mosquito netted cloth or stainless steel sieve.
6. Simmer this tomato juice, with the spices loosely tied in a cloth bag for about 20 min. in a covered vessel.
7. Then add vinegar and common salt.
8. Fill into hot sterilized 340 g bottles.
9. Seal them with crown corks. In boiling water for 2-3 min. before use.
10. Pasteurize bottles in boiling water for 30 min. Then cool and store them in cool, dry place.

TOMATO KETCHUP

Recipe:

Tomato pulp	:	3.0 kg
Onion chopped	:	37.0 g
Garlic chopped	:	2.5 g
Cloves, whole (headless)	:	1.0 g
Spices (coarsely powdered cardamom)	:	
Black pepper and cumin in equal quantities	:	1.5 g
Mace (Jalvatri), broken	:	0.25 g
Cinnamon	:	2.0 g
Red Chillies powder	:	2.0 g
Salt	:	3.0 g
Sugar	:	100.0 g
Vinegar	:	150.0 g
Or		
Glacial acetic acid	:	7.5 g
Sodium benzoate	:	0.75 g

Method:

1. Select fully ripe and red colour tomatoes.
2. Wash and cut them into small pieces.
3. Cook in a stainless steel or aluminium open pan till soft. Crush thoroughly with a wooden ladle while cooking.
4. Strain pulp through mosquito netted cloth or 1 mm mesh stainless steel sieve, by rubbing gently with bottom of enameled aluminium mug. Discard the seeds and skins.
5. Add 1/3 of the sugar and place spices in a muslin cloth bag and immerse it into the pulp.

6. Heat the pulp to about 1/3 of its original weight.

7. Remove the muslin cloth and squeeze.

8. Add vinegar, salt and sugar and again cook to about 1/3 of the original pulp.

9. Mix the preservative (Sodium benzoate) in small quantity of the finished product or water and transfer to the rest of the product.

10. Fill the ketchup into hot sterilized 340 g ketchup bottles.

11. Seal them with crown corks. Keep crown corks in boiling water for 2-3 min. before use.

12. Pasteurize bottles in boiling water for 30 min. Then cool and store them in cool, dry place till use.

TOMATO SOUP

Recipe:

Tomato pulp	:	3 kg
Water	:	1 kg
Onions (chopped)	:	50 kg
Garlic (chopped)	:	3 g
Salt	:	50 g
Butter	:	50 g
Sugar	:	60 g
Flour	:	30 g
Black per (Ground)	:	3 g
Cinnamon, Cardamom	:	
And other spices (as desired)	:	1.5 g

Method:

1. Select fully ripe and healthy red tomatoes.
2. Wash them thoroughly in fresh water.
3. Remove and discard the green and blemished portions.
4. Cut the fruit into small pieces and cook until soft.
5. Strain the pulp by rubbing through mosquito net cloth or stainless steel sieve to remove seeds and skin.
6. Boil the tomato pulp and add butter.
7. Add onions, garlic and water and simmer for 30 minutes.
8. Add salt, sugar, pepper and other spices and simmer for further 30 minutes.
9. Mix flour with a small quantity of water, boil and strain through cloth. Add strained liquid to a boiling tomato juice.

10. Pass the entire boiled mass (tomato soup) through a sieve or cloth with fine mesh.

11. Tomato soup is ready and may be served hot.

12. If tomato soup is required for later use and for long storage, then heat it again and fill hot into juice bottles. Seal the crown cork bottles and process for 45 min. at 115.5°C or 0.7 kg per cc. pressure in pressure cooker.

13 Cool the cans/bottles and store till use.

VALUE ADDITION OF MANGO

In general harvest maturity in mango is reached in 12-16 weeks after fruit set, depending on variety. Specific gravity is a good criterion to judge maturity of the fruit. Mangoes are harvested by hand if the pickers can reach them. Fruits on high branches are harvested with a picking pole having a cloth bag and cutting knife at the top. Fruits are to be harvested with little stalk to prevent latex trickling which leads to stem end rot. Mangoes are generally harvested at physiological mature stage and it takes 6-14 days to ripen under ambient conditions. The ripening phenomenon is associated with conversion of starch to sugars and loss of firmness of fruit. Mango is one such fruit, which can be processed at almost every stage of growth, development, maturity and ripening. Raw mango fruits are utilized for mango powder, pickle, chutney etc. An excellent drink can also be made from green mangoes. Ripe mangoes are utilized for making slab, toffee, various beverages such as nectar, squash etc. Drying after exposing to sulphur fumes also preserves ripe mango slices. Methods have also been standardized to produce cryogenically (liquid N) frozen mango slices. Mango fruits have been utilized for long time as fresh as well as its products like pickle, nectar, and squash. Pickles are good appetizers and add to the palatability of a meal. They stimulate the flow of gastric juice and thus help in digestion.

Green or ripe mangoes can be processed and value added into various products as follows:-

MANGO JAM

Ingredients:

Mango pulp	-	1 kg
Sugar	-	750 g
Citric acid	-	20 g

Method:

Mix the mango pulp and sugar thoroughly in a
clean stainless steel vessel

↓

Add citric acid and boil the mango pulp and sugar till jam
consistency (68°Brix) is reached.

↓

Fill hot in sterilized bottles, cool and store

MANGO PICKLE

Mangoes

(Mature, green)

↓

Washing

↓

Peeling

↓

Slicing

↓

Putting slices in jar

↓

Sprinkling salt

↓

Putting in sun for one week
(Shaking jar at least twice a day to mix the salt)

↓

Mixing spices

↓

Storage at ambient temperature
(in cool and dry place)

MANGO RTS

Fruit (Pulp / juice)

↓

Mixing with strained juice

↓

Preparation of syrup solution
(Sugar + water + acid, heated just to dissolve)

↓

Homogenization

↓

Bottling

↓

Crown corking

↓

Pasteurization
(at about 90°C) for 25 minutes

↓

Cooling

↓

Storage

SWEET MANGO CHUTNEY

Recipe:

Mango slices (or) shreds – 1.0 kg, sugar (or) gur – 1.0 kg, salt – 45 g, onions (chopped) – 50 g, garlic (chopped)-15 g , ginger (chopped) – 15 g, red chilli powder – 10 g, black pepper, cardamom, cinnamon, cumin – 10 g each, cloves – 5 nos. and vinegar – 170 ml.

Flow diagram for sweet mango chutney

Mature mangoes

↓

Washing

↓

Peeling

↓

Grating (or) slicing

↓

Cooking with a little water to make hightly soft

↓

Mixing with sugar and salt and leaving for an hour

Keeping all ingredients (except vinegar) in cloth bag, tied loosely, putting in mixture and cooking on low flame

During cooking spice bag pressed occasionally

Cooking to consistency of jam (up to 105°C) with stiming occasionally

Removal of spice bag after squeezing

Addition of vinegar

Cooking for 2-5 min.

Filling hot into bottles

Sealing (airtight)

Storage at ambient temperature

PROTEIN ENRICHED SPICY MANGO BAR

Method of preparation:

- Roast green gram dhal slightly to remove the raw flavour.
- Then ground into fine flour.
- Steam the green gram dhal flour for 10 minutes
- Dry it

- Pass through fine mesh to avoid lumps
- Steam the soy flour for 10 minutes to reduce the raw beany flavour and to inactivate trypsin inhibitor.
- Dry the steamed flour and pass it through fine mesh to avoid lumps
- Steam the soy protein isolate for 10 minutes.
- Dry the steamed flour
- Pass through fine mesh to avoid lumps

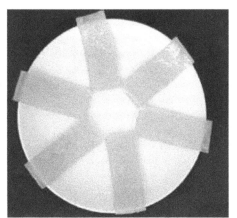

Preparation of spicy mango bar:

- Heat the mango pulp (80 g) for 3 minutes to inactivate enzymes
- Make a paste of corn flour (4 g) with little amount of water (to avoid the formation of lumps)
- Mix the pulp with corn flour paste, sugar (50 g), chilli powder (0.75 g), asafoetida (0.1 g) and green gram (20 g)/soy flour (20 g)/ soy protein isolate (10 g)/skim milk powder (10 g)
- Heat this mix on medium flame by stirring continuously using a wooden ladle up to a final TSS of 45°Brix
- Cool the concentrated pulp to room temperature and add 0.1% KMS and mix thoroughly. Spread the finished pulp evenly in an aluminum tray to a thickness of 0.5 cm
- Dry at 60°C for 6 hr in a cabinet drier

MANGO SLICES (AMCHUR)

Mangoes (mature, green)

↓

Washing

↓

Peeling

↓

Cutting into slices of suitable size

↓

Dipping in 2% salt solution for an hour

↓

Draining out and dipping in 2000 ppm SO_2 solution for 2 hr

↓

Spreading slices in thin layer on wooden trays

↓

Sun drying

↓

Filling into airtight containers

↓

Storage in dry place

MANGO CANDY

Ingredients:

Mango	:	1 kg
Sugar	:	1.120 kg
Water	:	500 ml
Citric acid	:	6.4 g
KMS	:	1.2 g

Method:

- Selection of ripe mango fruit
- Cutting into pieces
- Soaking in 2% $CaCl_2$ solution
- Preparation of sugar syrup (addition of sugar, water, citric acid)
- Cooking the fruit in sugar syrup till soft consistency (TSS 75°Brix and temp 106°C)
- Soaking the fruit for 7 days
- Boiling of sugar syrup to 60°Brix
- Packing the mango preserve in glass jar
- Draining the sugar syrup
- Drying in shade to get mango candy

FLOW CHART FOR PROCESSING OF FRUIT POWDER

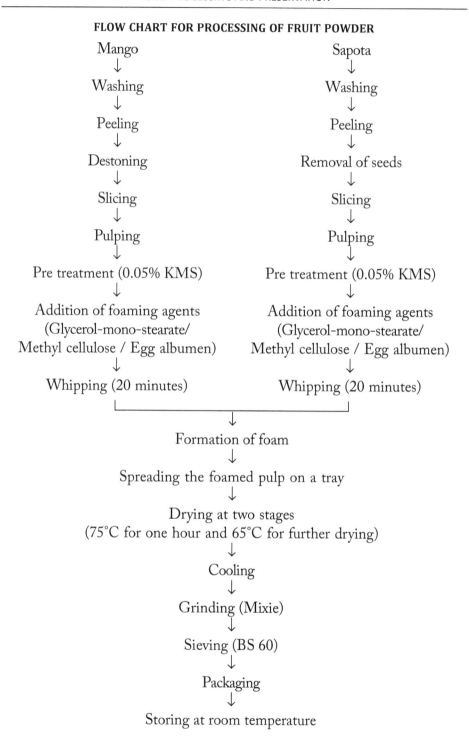

Mango	Sapota
↓	↓
Washing	Washing
↓	↓
Peeling	Peeling
↓	↓
Destoning	Removal of seeds
↓	↓
Slicing	Slicing
↓	↓
Pulping	Pulping
↓	↓
Pre treatment (0.05% KMS)	Pre treatment (0.05% KMS)
↓	↓
Addition of foaming agents (Glycerol-mono-stearate/ Methyl cellulose / Egg albumen)	Addition of foaming agents (Glycerol-mono-stearate/ Methyl cellulose / Egg albumen)
↓	↓
Whipping (20 minutes)	Whipping (20 minutes)

Formation of foam
↓
Spreading the foamed pulp on a tray
↓
Drying at two stages
(75°C for one hour and 65°C for further drying)
↓
Cooling
↓
Grinding (Mixie)
↓
Sieving (BS 60)
↓
Packaging
↓
Storing at room temperature

MANGO SQUASH

Selection of fruits

↓

Washing

↓

Pulping

↓

Preparation of syrup
(sugar + water + acid, heating just to dissolve)

↓

Straining

↓

Mixing with juice

↓

Addition of preservative
(0.6 g KMS / litre squash)

↓

Bottling

↓

Capping

↓

Storage

VALUE ADDITION OF APPLE

An apple is an edible fruit produced by an apple tree (*Malus domestica*). Apple trees are cultivated worldwide and are the most widely grown species in the genus *Malus*. The tree originated in Central Asia, where its wild ancestor, *Malus sieversii*, is still found today. Apples have been grown for thousands of years in Asia and Europe and were brought to North America by European colonists. Apples have religious and mythological significance in many cultures, including Norse, Greek, and European Christian tradition.

Apple trees are large if grown from seed. Generally, apple cultivars are propagated by grafting onto rootstocks, which control the size of the resulting tree. There are more than 7,500 known cultivars of apples, resulting in a range of desired characteristics. Different cultivars are bred for various tastes and use, including cooking, eating raw and cider production. Trees and fruit are prone to a number of fungal, bacterial and pest problems, which can be controlled by a number of organic and non-organic means. In 2010, the fruit's genome was sequenced as part of research on disease control and selective breeding in apple production. Worldwide production of apples in 2018 was 86 million tonnes, with China accounting for nearly half of the total.

DEHYDRATED APPLES

Turn on the dehydrator and set the temperature to 145°F

↓

Dry at this temperature for about one hour.

↓

Reduce the temperature to 135°F and finish dehydrating the apples until done, about 6-12 hours depending on the moisture level in the apples.

APPLE JUICE CONCENTRATE

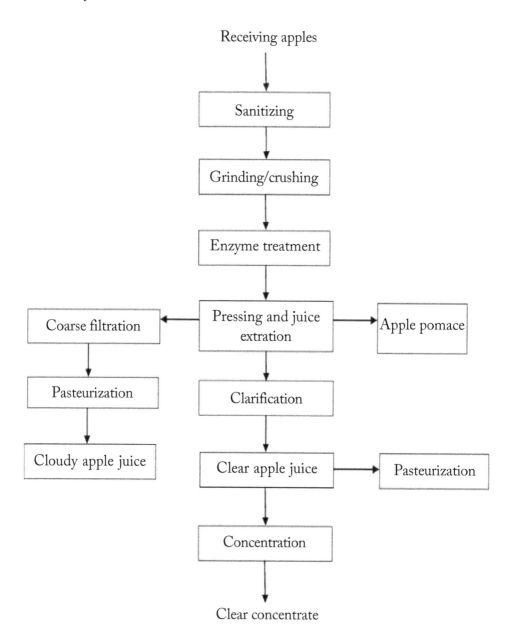

APPLE CIDER

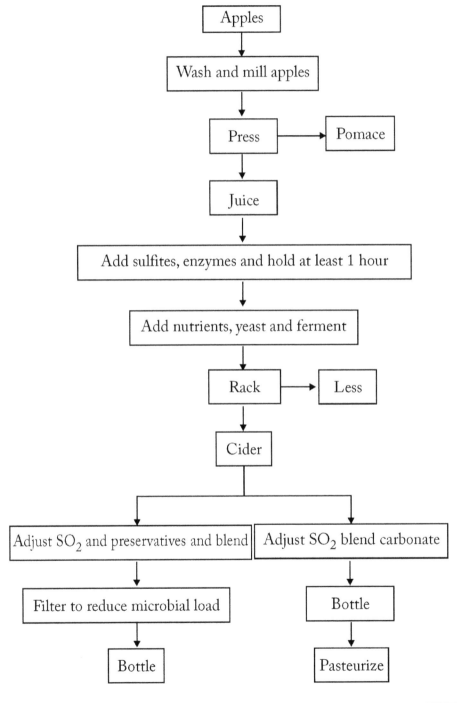

BOTTLING OF APPLES

Method:

1. Select medium size, fresh, firm, fully developed and sound apples.

2. Wash them thoroughly in running water to remove dust, spray residues etc.

3. Peel apples with s/s knives. Remove core and cut into slices or rings. Keep under salt water to avoid darkening (1tsp/1 lt. water)

4. Sterilize bottles (Pint size glass jars) by immersing them in cold water gradually heating to boiling point. Boil for 15 minutes.

5. Prepare 40% sugar syrup (1 cup of sugar to 1 cup of sugar to 1½ cups of water) by dissolving the sugar in boiling water and filtering through cloth.

6. Drain and rinse prepared apple slices or rings.

7. Fill into the containers. Pour boiling sugar syrup to cover the fruit.

8. Place the bottles into the boiling water in the open pan. The level of water in the pan should be slightly less (5cm) than the height of the container. To avoid bumping and breakage of bottles, place 3-4 folds of coarse cloth at the bottom of the pan.

9. Heat the water till the centre of the containers record temperature 89-82°C.

10. Seal the jars immediately while still hot, one by one with screw caps (screw caps with rubber rings).

11. Pasteurize (boil) the containers in boiling water for half an hour. Keep a thick pad of cloth at the bottom of the pan to avoid breakage of bottles (Jars).

12. Cool the jars in air and store in a cool and dry place till use.

APPLE GINGER JAM

Recipe:

Apples, prepared	:	3 kg
Root ginger, bruised, fresh	:	125 g
or		
Ginger, ground	:	25 g
Water	:	1 litre
Sugar	:	3 kg
Lime juice	:	60 ml (3-4 limes)
Or		
Citric /tartaric acid	:	1 level teaspoon (5g).

Method:

1. Peel, core and cut the apples into small pieces.
2. Place them with the water, acid (or lime juice) and ground ginger, if used in the cooking pan.
3. If root ginger is used, bruise it and put it in the muslin cloth in the pan.
4. Cook slowly until tender, and then remove the bag of ginger.
5. Add the sugar, stir until dissolved and boil rapidly until setting point is reached.
6. Pour Jam into hot jars and seal at once or let the jam cool and then pour a layer of hot melted paraffin wax before sealing.

APPLE JELLY

Method:

1. Select sound and mature fruits.

2. Wash and cut into $^{1/4}$inch thick slices. Cover the pieces with water and add citric 1 g /kg of fruit.

3. Boil for about half an hour and strain the extract through coarse cloth.

4. Take one more extract in a similar way, but add water half to the weight of fruit.

5. Mix both extracts and keep in a deep container overnight. Next day carefully decant the extract without disturbing it, and pass through jelly bag if available.

6. **Test for pectin:** Add two tea spoons of methylated or rectified spirit to one teaspoon of extract in a test tube or small container. Formation of one big clot indicates high pectin in the extract and thin gelatinous precipitate indicates poor pectin. When the extract is poor in pectin, concentrate till it gives test for high pectin.

7. Add ¾ or ½ kg of sugar for one kg of high or medium extract respectively. Boil till sugar dissolves and pass through muslin cloth.

8. Cook the mixture till it gives a perfect sheet test:

   Sheet Test: Dip a spoon in the product and allow it to fall down from the sides of spoon. If on cooling the product falls in the form of sheets and not in a steady stream, then the product is ready.

9. Pour the jelly into clean, dry glass jars. Allow the product to cool and add melted paraffin wax (1/2 inch thick) to the product. Seal air tight.

10. Store jars in a cool and dry place.

APPLE TOFFEE

Recipe:

Apple pulp	:	1 kg
Sugar	:	0.566 kg
Glucose	:	0.093 kg
Skimmed milk powder	:	0.140 kg
Hydrogenated fat	:	0.093 kg

Method:

1. Wash, peel and keep the fruit immersed in 0.5% potassium metabisulphite solution to check browning.
2. Core and cut apples into small pieces.
3. Pulp into fine mesh with the help of warning blender or cook till soft and pass through muslin netted cloth.
4. Cook pulp to 1/3 of its original volume.
5. Add fat and keep cooking.
6. Mix sugar and glucose and add slowly part to the mixture.
7. Make a thick paste of skimmed milk powder with a little quantity of water, heat and add to the mixture.
8. Concentrate the mass to such a final consistency when speck of a product does not break if put in a glass of water.
9. Spread on tray smeared with little fat or glycerine into thick slab of 1.5 inch thick.
10. Cut into small pieces and wrap in butter paper.
11. Keep the product in an air tight container.

APPLE FRUIT BAR

Flow chart:

Apples, ripe fruit (1kg)
↓
Washing

Peeling, coring, peel pomace and damaged fruit disposal (300 g)
↓
Soaking in solution (0.2% KMS+0.1% citric acid)
↓
Crushing / Grinding
↓
Pulping
↓
Pulp (700 g)
↓
Preserved pulp, mixed cultivars (14° Brix, acidity 0.3%)
↓
Adjustment to Brix (20°), sugar 42 g and acidity 0.3 % citric acid)
↓
Heating pulp till Brix 25°
↓
Partial cooling
↓
Spreading pulp on tray smeared with hydrogenated fat or glycerine
↓
Drying in oven under sun 4-5 days (40 hrs at 70°C)
↓
Cutting
↓
Packing in butter paper
↓

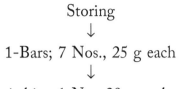

Storing

↓

1-Bars; 7 Nos., 25 g each

↓

2-fruit bits: 1 No., 20 g each pack

*KMS-Potassium metabisulphite

Method:

1. Select ripe but sound apples, preferably of mixed cultivars.

2. Peel, core and keep apples in KMS solution to check enzymatic browning.

3. Grate with s/s or plastic grater into five pieces.

4. Pulp in warming blender/mixer for ½ to 1 minute.

5. If required preserve pulp for long storage with KMS@ 0.3% in air tight jars. Dissolve KMS in small quantity of water and mix with pulp while blending.

6. Adjust Brix at 20°Brix and acidity at 0.5% of pulp by adding sugar and citric acid respectively

7. Heat pulp for 5-10 minutes till Brix reaching 25°.

8. Cool pulp.

9. Spread pulp on s/s or aluminium tray smeared with hydrogenated fat into ½ an inch thick layer.

10. Dry in oven for 3 hrs at 70°C till day.

11. Bar can also be dried under sun for 4 -5 days depending upon the temperature.

12. Cut apple fruit bars into suitable size with s/s knife and wrap in butter paper.

13. Keep wrapped bars in polythene bags and store in cool dry place.

APPLE PIE

Recipe:

Prepared apples	:	6 cups
(Thinly sliced)		
Sugar	:	¾ to 1 cup
All-purpose flour	:	2 tsp
Ground cinnamon	:	½ to 1 tsp
Ground nutmeg	:	1 dash
Salt	:	2 g
Plain pastry	:	2- crust 9 – inch pie
Butter	:	2 tsp

Method:

1. Combine sugar, flour, spices and dash salt mix with apples.
2. Line 9-inch pie plate with pastry and trim even with rim of pie plate.
3. Fill with apple mixture, dot with butter.
4. Adjust top crust. Trim upper top crust ½ inch beyond edge. Tuck (Seal) Top crust under edge of lower crust.
5. Cut slits in top crust for escape of steam.
6. Sprinkle with sugar.
7. Bake at 400°F (205°C) for 50 minutes or till done.

FOR SPARING

PLAIN PASTRY FOR APPLE PIE:

(Two 9-inch single –crust pie)

Recipe:

All –purpose flour	:	2 cups
Salt	:	1 tsp
Shortening	:	2/3 cup
Water	:	5-7 tsp

Method:

1. Sift flour and salt together.
2. Cut in shortening with mixture into small pieces.
3. Sprinkle one tablespoon of cold water over part of the mixture and gently toss with fork. Push to side of bowl. Repeat till all mixture is moistened.
4. Form into two bowls.
5. Flatten flour ball one by one on lightly floured surface by pressing with edge of hand three times across in both directions.
6. Roll from centre to edge till 1/8 inch thick and pie crusts are ready for apple pie.

APPLE –TOMATO CHUTENEY

Recipe:

Prepared apples	:	3 kg
Prepared tomatoes	:	2 kg
Onions chopped	:	0.5 kg
Sugar	:	3.5 kg
Raisin, optional	:	75 g
Salt	:	150 g
Ginger, dried	:	150 g
Red chilly powder	:	10 g
Cinnamon	:	5 g
Cardamom	:	5 g
Clove	:	2 g
Black pepper	:	5 g
Vinegar	:	1 litre
Glacial acetic acid	:	40 ml

Method:

1. Select fully ripe, red tomatoes and wash them in fresh water.

2. Peel tomatoes by dipping them in boiling water for three minutes and then immersing in cold water. This helps in removal of the skin. Cut and crush the peeled tomatoes.

3. Peel, core and slice the apples into thin pieces with the help of grater.

4. Cook all the ingredients except vinegar to a thick consistency.

5. Add vinegar, cook again till thick and pack the chutney while hot in wide mouthed dry bottles.

6. Seal the Jars air tight and store in a cool, dry place.

VALUE ADDITION OF BANANA

Asia produces about 40% of the banana out of global annual production of 45 million metric tones. With the available technologies, post-harvest losses can be reduced to half and simultaneously value addition can be made for this produce. An attempt is made to pool the knowledge of post-harvest handling to improve the shelf life, processing and by-products utilization of banana crop for producing value added products from it. The harvesting methods, techniques for prolonging the shelf life of fruit using chemical dips, low-temperature methods, controlled atmosphere storage, sub atmospheric-pressure storage, etc. are discussed. The banana is a versatile fruit for preparing several processed foods through simple processing methods. Processing techniques of several products such as pulp, juice, canned slices, jams, deep-fat-fried chips (crisps), toffee, fig, fruit bars, brandy, etc. are presented. Apart from these, value addition techniques for utilization of by-products are also emphasized.

PROCESSING

In general, to obtain a good quality product from ripe bananas the fruit is harvested green and ripened artificially under controlled conditions at the processing factory. The banana is a versatile fruit for preparing several processed foods through simple processing methods. Ripe and unripe bananas can be successfully processed into several products such as pulp, juice, canned slices, jams, deep-fat-fried chips (crisps), toffee, fig, fruit bars, brandy, etc. Banana products can be divided into two types – those for direct consumption such as figs and those for use in the food manufacturing industry, for example purées and powder.

PULP, JUICE AND CONCENTRATE

Fully ripe fruits are washed, peeled and forced through a screw type pulper. Then the pulp is homogenized, deaerated using a centrifugal deaerator and held under vacuum. This method eliminates the need for steam blanching, which may be responsible for oxidative colour and flavour changes. The pulp contains leuko-anthocyanins, which in the conventional method of canning such as low acid food cause pink discolouration. The homogenized deaerated pulp is heated, cooled and filled aseptically into containers. For preparation of clarified juice, the banana pulp is mixed with an appropriate level of SO_2 and treated with pectic enzyme at 40°C until the clear juice is separated from the pomace. The enzyme dosage of 0.75-1% (V/W of pulp) is optimum. The mass is passed through cheesecloth and filtered using a filter aid like infusorial-earth such as Hy-flow Super Cel or Dicalite. The filtered juice is blended with sugar and acid, if needed, pasteurized, filled hot into pasteurized bottles, sealed and cooled. Banana juice clarified by pectolytic enzyme treatment can be made into concentrate using vacuum evaporators. Aroma recovery, concentration and addition to the concentrated juice are essential to obtain fullflavoured concentrate.

BANANA PULP

Fully ripe fruits are washed, peeled and forced through a screw type pulper.
↓
The pulp is homogenized, deaerated using a centrifugal deaerator and held under vacuum.
↓
Homogenized deaerated pulp is heated, cooled and filled aseptically into containers.

Preparation of Clarified Juice and Concentrate

Banana pulp is mixed with an appropriate level of SO_2 and treated with pectic enzyme at 40°C until the clear juice is separated from the pomace.

↓

The enzyme dosage of 0.75-1% (V/W of pulp) is optimum.

↓

The mass is passed through cheesecloth and filtered using a filter aid like infusorial-earth such as Hy-flow Super Cel or Dicalite.

↓

The filtered juice is blended with sugar and acid, if needed, pasteurized, filled hot into pasteurized bottles, sealed and cooled.

↓

Banana juice clarified by pectolytic enzyme treatment can be made into concentrate using vacuum evaporators.

↓

Aroma recovery, concentration and addition to the concentrated juice are essential to obtain fullflavoured concentrate.

BANANA TOFFEE

In the production of toffee, banana pulp is concentrated in a steam-jacketed kettle to about one-third of its original volume. Other ingredients, namely, sugar, glucose, skim milk power and *vanaspati* (hydrogenated oil) are added, mixed and the cooking continued to a final weight equal to about 20% of the fruit pulp taken. The cooked mass is transferred to a smooth level surface and smeared to a thin sheet. It is allowed to cool and set for 2 h. The solid sheet is cut in to pieces and dried at 50-55°C to a final moisture content of 5-6%, after which the pieces are wrapped and stored/marketed.

Preparation of Banana Toffee

Banana pulp is concentrated in a steam-jacketed kettle to about one-third of its original volume.

Other ingredients, namely, sugar, glucose, skim milk powder and vanaspati (hydrogenated oil) are added, mixed and the cooking continued to a final weight equal to about 20% of the fruit pulp taken.

The cooked mass is transferred to a smooth level surface and smeared to a thin sheet.

Allow to cool and set for 2 h.

The solid sheet is cut in to pieces and dried at 50-55°C to a final moisture content of 5-6%

After which the pieces are wrapped and stored/marketed.

BANANA JAM

The jam is made by boiling equal quantities of fruit and sugar together with water and lemon juice, lime juice of citric acid, until settling point is reached.

Dried Slices/Figs Dried slices

These are prepared from ripe fruits with a sugar content of about 19.5%. Banana figs, or fingers as they are sometimes known, are usually whole, peeled fruit carefully dried so as to retain their shape, although sometimes the fruit is sliced or halved lengthwise to facilitate dying. After peeling, these are treated with sulphurous acid and then dried as soon as possible. Drying is done either in the sun or in a dehydrator at controlled temperature and relative humidity. The peeled whole or sliced or halved fruits are spread on wooden trays made of slats and sulfured for an hour by exposure to burning sulfur. The slices are then dried at 55-60°C in a cabinet dryer for 20 h so that the product is pliable, soft and non-sticky. There are various drying systems using temperatures between 50 and 82°C for 10-24 h to give moisture content ranging from 8-18% and a yield of dried figs of 12 to 17% of the fresh banana. Bananas can also be dried by osmotic dehydration, using a technique which involves drawing water from ¼" thick banana slices by placing them in a sugar solution of 67-70° Brix for 8-10 h to reduce 50% of its original weight, followed by vacuum drying at 65-70°C at a vacuum of 10 mm Hg for 5 h. The moisture content of the final product is 2.5% or less, much lower than that achieved by other methods. Dried product is packed in cardboard cartons lined with polyethylene film. The major problem with this product is browning during storage. Banana figs can be stored for one year. Freeze-dried banana slices with better organoleptic properties than air-dried products can be prepared from ripe fruits. Freeze-dried slices give more acceptable banana powder than air-dried slices.

Preparation:

Ripe fruits with a sugar content of about 19.5%.

Banana figs, or fingers as they are sometimes known, are usually whole, peeled fruit carefully dried so as these are treated with sulphurous acid and then dried as soon as possible.

Drying is done either in the sun or in a dehydrator at controlled temperature and relative humidity.

The peeled whole or sliced or halved fruits are spread on wooden trays made of slats and sulfured for an hour by exposure to burning sulfur.

The slices are then dried at 55-60°C in a cabinet dryer for 20 h so that the product is pliable, soft and non-sticky.

↓

There are various drying systems using temperatures between 50 and 82°C for 10-24 h to give moisture content ranging from 8-18% and a yield of dried figs of 12 to 17% of the fresh banana.

OSMOTIC DEHYDRATED BANANA SLICES:

Bananas can also be dried by osmotic dehydration, using a technique which involves drawing water from ¼" thick banana slices by placing them in a sugar solution of 67-70° Brix for 8-10 h to reduce 50% of its original weight, followed by vacuum drying at 65-70°C at a vacuum of 10 mm Hg for 5 h. The moisture content of the final product is 2.5% or less, much lower than that achieved by other methods. Dried product is packed in cardboard cartons lined with polyethylene film. The major problem with this product is browning during storage. Banana figs can be stored for one year. Freeze-dried banana slices with better organoleptic properties than air-dried products can be prepared from ripe fruits. Freeze-dried slices give more acceptable banana powder than air-dried slices.

BANANA CHIPS/CRISPS

The Nendran plantain banana is famous for preparation of chips. In the production of banana and plantain chips (crisps), thin banana slices of 1.75-2.0 mm thick are soaked in a solution containing NaCl, citric acid and potassium metabisulfite for 30 min. They are removed, wiped and fried in hydrogenated fat or edible oil at 180-200°C. The frying medium:material ratio should be about 4:1. Common salt at the rate of 0.6% (W/W added as 20% aqueous solution) is sprinkled over the frying chips in the pan towards the end of frying. The excess oil is removed by wiping and dusted with salt, if previous process is not followed. Antioxidants are added to keep the product free from becoming rancid. Addition of 0.1-0.15% turmeric powder during frying enhance the colour. Alternatively, slices may be dried before frying and the antioxidant and salt added with the oil. The chips are stored in 300 gauge high-density polyethylene bags or 400 gauge low-density polyethylene bags, which in turn are kept in sealed tins under CO_2. They remain good for 6 months. They are eaten as snack like potato chips.

Preparation:

Thin banana slices of 1.75-2.0 mm thick are soaked in a solution containing NaCl, citric acid and potassium metabisulfite for 30 min.

They are removed, wiped and fried in hydrogenated fat or edible oil at 180-200°C.

The frying medium:material ratio should be about 4:1.

Common salt at the rate of 0.6% (W/W added as 20% aqueous solution) is sprinkled over the frying chips in the pan towards the end of frying.

The excess oil is removed by wiping and dusted with salt, if previous process is not followed. Antioxidants are added to keep the product free from becoming rancid.

Addition of 0.1-0.15% turmeric powder during frying enhance the colour.

Slices may be dried before frying and the antioxidant and salt added with the oil.

↓

The chips are stored in 300 gauge high-density polyethylene bags or 400 gauge low-density polyethylene bags, which in turn are kept in sealed tins under CO_2.

BANANA PUREE

Banana purée is by far the most important processed product made from the pulp of ripe fruit. Preserving the pulp is done by one of the three methods: canning aseptically, acidification followed by normal canning or quick freezing. Peeled, ripe fruits are conveyed to a pump which forces them through a plate with ¼" holes, then onto a homogeniser, followed by a centrifugal deaerator and into a receiving tank with 29" vacuum, where the removal of air helps to prevent discolouration by oxidation. The purée is passed through a series of scrapped surface heat exchanger where it is sterilized by steam, partially cooled and finally brought to filling temperature. The sterilized purée is then packed aseptically into steam-sterilized cans, which are closed in a steam atmosphere. The purée is used as an ingredient in dairy desserts, bakery items, drinks, processed foods and sauces and as part of special diets in hospitals and nursing homes. Ripe bananas are also sliced and canned in an acidified syrup and are used in desserts, fruit salads, cocktail drinks, baby foods and bakery items.

Vinegar from banana pulp and peel

Vinegar has been prepared from a mash consisting of the pulp and peel of the ripe fruit. The mash is pasteurized and then inoculated with a pure culture of *Saccharomyces ellipsoids*. Fermentation is allowed to proceed at 20 to 23°C for 14-20 days. The alcohol content of mash amounts to 6.55 to 10.12%. The 'cider' thus obtained is filtered and subjected to acidification giving a yield of vinegar of 48-53% based on the weight of the fruit taken.

Preparation of Vinegar:

Vinegar has been prepared from a mash consisting of the pulp and peel of the ripe fruit.

The mash is pasteurized and then inoculated with a pure culture of *Saccharomyces ellipsoids*

Fermentation is allowed to proceed at 20 to 23°C for 14-20 days.

The alcohol content of mash amounts to 6.55 to 10.12%.

The 'cider' thus obtained is filtered and subjected to acidification giving a yield of vinegar of 48-53% based on the weight of the fruit taken.

VALUE ADDITION OF GRAPES

A grape is a fruit, botanically a berry, of the deciduous woody vines of the flowering plant genus *Vitis*. Grapes can be eaten fresh as table grapes or they can be used for making wine, jam, grape juice, jelly, grape seed extract, raisins, vinegar, and grape seed oil. Grapes are a non-climacteric type of fruit, generally occurring in clusters. Raw grapes are 81% water, 18% carbohydrates, 1% protein, and have negligible fat (table). A 100-gram (3½-ounce) reference amount of raw grapes supplies 288 kilojoules (69 kilocalories) of food energy and a moderate amount of vitamin K (14% of the Daily Value), with no other micronutrients in significant content.

GRAPE JUICE

One of the processed products of grape includes grape juice. It is very refreshing and healthy at the same time. Processing of grapes into juice is not that simple task; it needs expertise to get the right flavor, taste and color. The quality of a juice depends upon the sugar level, acid content and flavor constituents. The composition of a juice varies from year to year and changes during ripening. Several factors which influence the composition of a juice are soil, climatic conditions, biotic, abiotic stress and vineyard management practices. Grape juice gets its color from the anthocyanin pigments located in or near the skin. Grape juice is usually extracted using a hot-press or a coldpress technique and the temperature have a significant effect on its quality. Hot press techniques yields more juice than cold press. Increased extraction temperature increases the color of the juice. To obtain a good color and flavor, the grapes are crushed and mixed with polygalacturonase and SO_2, followed by holding it for 24hr at room temperature before low temperature extraction. Chilled, ready to serve drinks have special packaging requirements. For more than 20 years the preferred and best packaging material for these kind of juices are paper board cartons. It tastes great and is safe and healthy to consume. Generally these kinds of drinks are refrigerated all along. They are best when served chill.

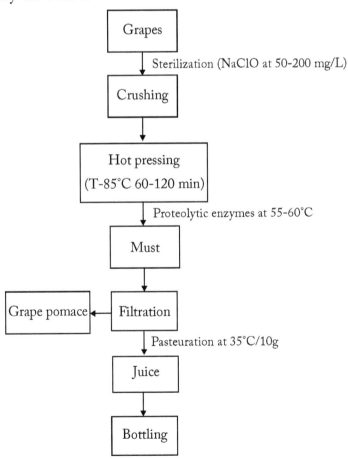

Vinegar from grapes

In French, Vinegar means sour ("aigre") wine ("vin"). Vinegars can be produced from various raw materials. Converting a good wine to vinegar might be considered as waste by someone but considering the economic reasons this could also be a profitable plan. High quality vinegars are most preferred than the wines from which they are produced. Grapes (*Vitis vinifera*) belong to the Vitaceae family and are amongst the largest cultivated global fruit crop due to their excessive use as table fruit and winemaking. As fresh fruit, grapes are very delicate and perishable, so the loss at harvest and during the distribution is very high. Due to grapes' delicateness and extreme perishability, the losses suffered during the preparation, harvest, packing, storing, transport and distribution of table grapes are very high. Post-harvest losses are mainly due to shattering of berries and grey mould. The value addition in grapes and diversification of grape products provide good returns to the growers as

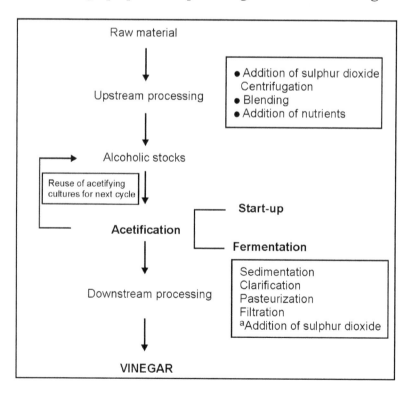

well as reduces the post-harvest losses. Grapes are utilized as raw material mainly for preparation of raisins, wine making, vinegar and juice concentrates and vinegar etc.

GRAPE WINE

Wine is one of the major processed products of grapes. Fermentation is the process through which wine is produced. This process is initiated by adding yeast with must (red wine) or juice (white wine). During fermentation, sugar contained in the must/juice is transformed to alcohol with output of carbon dioxide. Between 4°C and 36°C, yeast fulfills its task and the process stops when the sugar transforms completely. The process can even be stopped artificially by lowering the tank temperature. Wine gets its color from the grape skins and similarly rose wine and white wine can be made from the juice of red grapes and white grapes respectively. For making red and white wine, the suitable temperature for fermentation is 15-17 and 18-22°C and it's usually stored in steel or oak containers. The fermentation process lasts only a few days for lighter wines and up to 30 days for stronger wines. Based on the contact of the juice with the skins, the color and the tannin content gets stronger. This gives a longer lifespan for the wine however too much of tannins can ruin the wine. Red wines are separated from the skins and seeds and then their aging process starts in barrels and later bottles. Since white wine are fermented without maceration, white wines differ from red wines in terms of color and has lower tannin content, lighter body, a higher acidity and short aging time. The way the wine has been packaged has not changed much, till now. But in the last two decades, manufacturers have introduced new packaging methods to pack wines. The packaging includes pet bottles, bag in box, tetra pak, and glass bottles. These glass bottles are not the general one what we have come across, they are modified and made in different shapes and colors and at the same time made in lighter weight.

RAISINS

Dried grapes are otherwise known as raisins. It is one of the end products of grapes, we can't categorize this under processed but at the same time we can't say it is not processed, so it is considered as a semi processed grape product. About 90% of raisins produced globally are made by drying Thompson Seedless grapes and its clones. The grape bunches are treated with grape drying oil before drying. This oil is prepared by adding 15ml ethyl oleate and 25 g potassium carbonate in one liter of water. Various countries make use of "Drying on Vine (DOV)" practice for grape drying. While drying, water in the grape berry moves to the cuticle through the cells and then passes as vapor through the wax platelets and escapes from the outer surface. For faster drying of grapes high air temperature, rapid air movement and lower relative humidity are the climatic conditions required. For larger berries and thicker skins, the drying time is much higher. Green colored raisins fetch more price than the dark brown colored. The color depends on the drying temperature, presence of sunlight and humidity. Raisins with moisture content of 12-15% are well suited for storage. During storage, low moisture raisins become hard and high moisture raisins loses its quality. Raisins are generally packed in plastic bags, in bulk quantities it is packed in cartons or tins. Usually when you pick raisins from a super market it will be packed in a plastic bag.

VALUE ADDITION OF ORANGE

Pakistan ranks among top ten citrus producing and exporting countries of the world with an annual production of 1.7 million tonnes from an area of 651.8 thousand hectares. Out of total production 95% is contributed from a single province i.e. Punjab and Sargodha division is sharing almost 25%. The major part of our produce is exported after waxing as fresh commodity for dessert purpose and only a very limited quantity is processed in the form of fresh juice and other valuable products. There is a lot of potential to generate diversified high value products from the Kinnow mandarin and waste material produced after juice extraction. In this paper a large number of citrus products are suggested for possible use of the kinnow fruit to produce high value products for good returns to the farming community as well as sustainability of citrus industry of Pakistan.

ORANGE JUICE

Orange juice is a liquid extract of the orange tree fruit, produced by squeezing or reaming oranges. It comes in several different varieties, including blood orange, navel oranges, valencia orange, clementine, and tangerine.

As well as variations in oranges used, some varieties include differing amounts of juice vesicles, known as "pulp" in American English, and "(juicy) bits" in British English. These vesicles contain the juice of the orange and can be left in or removed during the manufacturing process. How juicy these vesicles are depend upon many factors, such as species, variety, and season. In American English, the beverage name is often abbreviated as "OJ". Commercial orange juice with a long shelf life is made by pasteurizing the juice and removing the oxygen from it. This removes much of the taste, necessitating the later addition of a flavor pack, generally made from orange products. Additionally, some juice is further processed by drying and later rehydrating the juice, or by concentrating the juice and later adding water to the concentrate. The health value of orange juice is debatable: it has a high concentration of vitamin C, but also a very high concentration of simple sugars, comparable to soft drinks. As a result, some government nutritional advice has been adjusted to encourage substitution of orange juice with raw fruit, which is digested more slowly, and limit daily consumption.

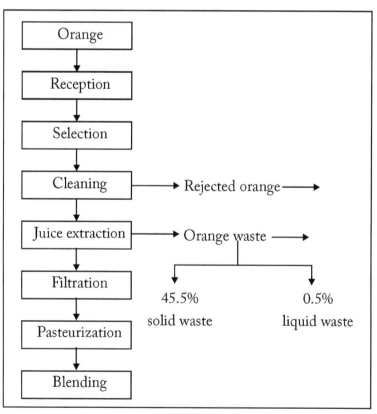

ORANGE SQUASH

Recipe:

Orange Juice	:	1 kg
Sugar	:	1.4 kg
Water	:	0.6 kg
Citric acid	:	37 g
Edible Orange colour	:	as per desired shade
Orange essence	:	few drops
Potassium metabisulphite	:	2 g

(Recipe worked out on the basis of final product as 33-34% juice, 50 brix 1.5% acidity, when orange juice having 10°Brix and 0.8% acidity)

Method:

1. Select fully ripe and coloured fruits.
2. Wash them with fresh water. Remove the peels and extract juice in screw press or any other suitable orange juice extractor.
3. Strain the juice through a coarse muslin cloth.
4. Prepare syrup by heating the mixture of sugar, acid and water. Strain it through a coarse muslin cloth.
5. Cool syrup and mix with orange juice
6. Add colour after first dissolving it in small quantity of squash or water.
7. Then add orange essence or oil. At home peel emulsion may be added which is prepared by crushing the peels of orange (2-4) oranges for every 100 fruits in a pestle motor and squeezing the crushed pulp through a thick cloth.
8. Add potassium metabisulphite after first dissolving it in small quantity of water.

9.   Fill the product into clean bottles. Screw tight the lids and seal them by dipping the top in melted paraffin wax.

10. Store in a cool and dry place till use.

Orange Syrup

The fruits are picked when fully ripe. After washing, the oranges are peeled and cut in quarters. With a manual extractor, the juice is separated from the skin and the pips. The juice is filtered through a fine mesh strainer and mixed with sugar at 50% of its weight. Heating to boiling point will allow the stabilization of micro-organisms which cause fermentation; any excess essential oil can be removed by skimming. Bottling and capping must be done while the juice is hot. Cooling, cleaning and labeling are the last operations before storage.

Delivery → Sorting → Washing → Peeling → Cutting → Extraction → Filtration → Mixing → Bottling → Capping → Sterilization → Cooling → Cleaning → Labeling → Storage

ORANGE MARMALADE

All types of oranges like santaras (loose skinned oranges), maltas (tight skinned oranges) and khattas (rough lemons) are suitable for making orange marmalade. Sound fruits and even of inferior quality having very little of juice, but otherwise sound can be used. Take two oranges either maltas or santaras for each khattas.

Method:

1.   Wash the fruits in fresh cold water. Remove the skin of santaras; and in the case of maltas, peel off the outer portion with a s/s knife retaining as much of albedo (white portion) portion as possible.

2.   Cut the peels of some orange into fine shreds and boil them in sufficient water for about 10 minutes. Change the water 2 to 3 times.

3. Cut the fruit into thin slices and cover with enough water in a pan (water equal quantity of fruit).

4. Cook the mass for about half an hour. Crush the fruit while cooking with a wooden laddle.

5. Strain the cooked mass through coarse muslin cloth.

6. Add water to 1/4 to the quantity of pomace and take a second and third extraction, if necessary.

7. Mix the two or three extract and keep the mixture in a deep vessel for sedimentation.

8. Siphon off the clear extract and discard the residual pulp.

9. **Pectin test:** Add two tea spoons of methylated or rectified spirit to one teaspoon of extract in a test tube or small container. Formation of one big colt indicates high pectin in the extract and thin gelatinous precipitate indicates poor pectin. When the extract is poor in pectin, concentrate by heating till it gives for high pectin.

10. Add ¾ to ½ parts of sugar for every one part of high or medium pectin extract and cook the mixture.

11. Add the boiled shreds of the orange peel while cooking.

12. Continue boiling till the mixture becomes thick or the product falls from the ladle in the form of flakes or sheets.

13. Fill the orange marmalade into clean, dry glass jars and seal them air tight after cooling.

ESSENTIAL OILS

In contrast to most essential oils, it is extracted as a by-product of orange juice production by centrifugation, producing cold-pressed oil. It is composed of mostly (greater than 90%) d-limonene, and is often used in place of pure d-limonene. D-limonene can be extracted from the oil by distillation.

Improved Steam Distillation (ISD)

The water washed orange peels were pre-heated in a vessel for about 30 minutes at a temperature of 50°C. Fresh orange peels are taken and preheated at a temperature below 50°C for about 30 minutes prior to steam distillation procedure. A little amount of distilled water that is enough to immerse the orange peels is added. The pre-heated sample was then loaded into the round bottomed flask and two-third of it was filled with the distilled water including the water that retained after preheating the cut orange peels. The condenser water was turned on and heated up to 100°C for two hours. Water loss occurred due to continuous heating. If the water level got too low, the sugar present in oranges would caramelize and burn. To avoid this two-third of the round bottomed flask was filled with water. The distillation was then run for about 120 minutes, at varied temperature ranges (10°C, 20°C and 30°C) of the heating mantle. The distillate was collected and the presence of the orange oil in the condensate was confirmed by the cloudy appearance on the top of the distillate.

Essential oils have a vast range of uses, ranging from the use in cleaning products, helping improve skin conditions and fighting illness. One of the most popular of these oils is sweet orange which has a wide array of natural health benefits as follows.

1.   Improves digestion and helps to relieve constipation.

2.   Nourishes dry, irritated and acne-prone skin when mixed with carrier oil and applied as a cream.

3.   Promotes a feeling of happiness and warmth when used in aromatherapy. Helps to eliminate toxins from the body.

4.   Helps in stimulating lymphatic action to promote balance in water processes and detoxification of the body.

5.   Contains all-natural antimicrobial properties. In one 2009 study published in the International Journal of Food and Science Technology, Orange oil was found to prevent the proliferation of *Escherichia coli* bacteria.

6.   Natural remedy for high blood pressure.

7.   Has anti-inflammatory properties. Sweet orange is one of several popular anti-inflammatory oils, including lemon, pine and eucalyptus. Orange oil has been shown to be the most effective.

8.   Reduces anxiety and boosts mood.

9.   Natural anti-depressant. Orange oil has been used as a mild tranquilizer for centuries. As little as five minutes of exposure to diffused orange oil can alter moods and enhance motivation, relaxation and clarity.

SUN DRYING OF PEARS

Flow Chart:

Pears (local cv. Sarkoh Tang)

↓

Washing

↓

Cutting into halves and removing cores

↓

Keeping fruit in salt water

↓

Draining, washing and soaking in 0.5% KMS* sol. For 2 hrs.

↓

Draining and spreading on wooden trays (single layers)

↓

Drying in sun for 6-7 days

↓

Keeping dried fruit in container for 1-2 days
(equalization of moisture)

↓

Packaging in polyethylene bags

*KMS: Potassium metabisulphite

Method:

1. Select fully ripe (but not soft) pear.
2. Wash the fruit thoroughly and remove with fresh water
3. Cut in half lengthwise and remove the core with spoon. . Keep the cut and cored fruit in salt water (1-2 per cent) to check browning).
4. Drain, wash and soak the prepared fruit in 0.5% KMS for two hours

PLUM SQUASH

Recipe:

Plum juice	:	1 lit
Sugar	:	2 kg
Water	:	1 Lit
Citric Acid	:	10 g
Sodium Benzoate	:	4 g

Method:

1. Take 1 litre of water in a patila
2. Boil on a heating source
3. Add sugar ad Citric Acid to make syrup
4. Cool
5. Add juice with sugar syrup and mix thoroughly
6. Add preservative
7. Bottle it and cap properly

JAMS, JELLIES AND MARMALADES

JAMS

These are solid gels made from fruit pulp or juice, sugar and added pectin. They can be made from single fruits or a combination of fruits. The fruit content should be at least 40%. In mixed fruit jams the first-named fruit should be at least 50% of the total fruit added (based on UK legislation). The total sugar content of jam should not be less than 68%.

JELLIES

These are crystal clear jams, produced using filtered juice instead of fruit pulp.

MARMALADES

These are produced mainly from clear citrus juices. They have fine shreds of peel suspended in the gel. Commonly used fruits include lime, orange, grapefruit, lemon and orange. Ginger may also be used alone or in combination with these citrus fruits. The fruit content should not be less than 20% citrus fruit and the sugar content is similar to jams.

Fruit preparation:

Fruit should be sorted and cleaned thoroughly. Only mature fruit, without mould, excessive bruising or insect damage should be used. All stems, leaves, skins etc. should be removed.

Ingredient mixing:

Accurate scales are needed to weigh out the ingredients and care is needed to make sure that the correct weights are used each time. In particular pectin powder should be thoroughly mixed with sugar to prevent lumps forming and resulting in a weak gel.

Fruit pulp/juice:

It is possible, by hand, to peel and pulp the fruit, press and filter the juice but this level of production is very low (e.g. 10-20 half kilogram jars per day) and the procedure is both laborious and time consuming. For small-scale commercial production it is better to use small manual or powered equipment to pulp the fruit and/or express the juice. Juice or pulp contains enzymes and in many fruits these cause rapid browning if they are not destroyed or inhibited from acting. The easiest way to control browning is to heat the juice in small batches as it is produced, rather than producing a large amount and storing it for several hours before use. The procedure described under 'batch preparation' and 'boiling' has been found to work very well.

Sugar:

Refined, granular, white sugar should be used whenever possible, but even these will often contain small amounts of material (e.g. black specks) which reduce the value of a preserve. The sugar should therefore be dissolved in water to make a strong syrup and then filtered through muslin cloth or a fine mesh. It is most important that the filters and pans are thoroughly cleaned each day to prevent insects and micro-organisms from contaminating the equipment.

The strength of the sugar syrup can be easily calculated as follows:

$$\text{Percent sugar} = \frac{\text{Weight of sugar}}{\text{Weight of sugar} + \text{Weight of water}} \times 100$$

So for example a 50% sugar solution (50°Brix) could be made by dissolving 500g sugar in 500 mL water.

Pectin:

All fruits contain pectin in the skins and to a lesser extent in the pulp. However, the amount of pectin varies with the type of fruit and the stage of maturity. Apples, citrus peels and passion fruit for example, contain a high concentration of pectin; strawberries and melon contain less. In general, the amount of pectin in fruit decreases as the fruit matures.

Although it is possible to get a good preserve using the pectin already in the fruit, it is better to buy pectin powder or solution and add a known amount to the fruit juice or pulp. This will produce a standardized gel each time and there will be less risk of a batch failing to set.

There are many different types of pectin available, but for preserves, a slow setting type is needed to allow the gel to form in the jar during cooling. However, in larger containers (e.g. 5-10 kg jars of jam) or for preserves in which peels or pieces of fruit are suspended in the gel, a faster setting pectin in needed. In both types, the concentration of pectin varies from 0.2-0.7% depending on the type of fruit being used. Pectin is usually supplied as '150 grade' (or 150 SAG) which indicates the ratio of the weight of sugar to weight of pectin that will produce a standard strength of gel when the preserve is boiled to 65% soluble solids. 5 SAG is normally enough to produce a good gel.

An example of a calculation to find the weight of pectin to be added is as follows:

150 SAG pectin is diluted to 5 SAG, i.e. a 30 fold dilution. Therefore 3.3 g pectin would be used for every 100 g of material.

However, if commercially produced pectin cannot be obtained it is possible to produce a pectin solution by boiling the sliced skins of passion fruit, lime, lemon, orange or grapefruit in water for 20-30 minutes. The solution should be filtered before adding to the fruit pulp. The amount of solution to be added depends on the type of fruit and a number of other factors, and must be found by trial and error.

Acid:

Acids are added to fruit juice to bring the pH within the range 3.0-3.3 which is necessary for jam making (pH is a measure of acidity - lower pH means greater acidity). As the acidity varies in different types of fruit and also in different samples of the same fruit, it may be necessary to check for the correct acidity if different fruits are used. (*NB limes have a lower pH than 3.3 and sodium bicarbonate is needed to reduce the acidity*).

The only acids that are allowed to be added to jam are *citric acid, tartaric acid and malic acid*. In practice citric acid is usually used and this is widely available from chemists or pharmacies.

Food colours:

Some fruit pulps/juices do not substantially change colour during boiling and in others the colour change is acceptable. In both cases it is not necessary to add artificial colours. However, some fruits become dark brown and are not sufficiently attractive to customers. In these cases, small quantities of permitted colours may be added if no other fruits are available for use. The list of permitted colours differ throughout the world and it is necessary to check with the local Bureau of Standards to see which colours are allowed.

Batch preparation:

First thoroughly mix pectin powder with 5 times its weight of sugar, this will allow the pectin to fully dissolve without forming lumps. The amount of sugar, pectin, fruit pulp/juice and acid needed will depend on the type of fruit and the customers' requirements. However, as an example of a typical product, the following recipe has been successfully used to make water melon jam:

115 Kg water melon, 55 Kg sugar, 0.9 Kg ginger, 0.47 Kg citric acid, 0.66 Kg pectin. Mix together the sugar/pectin, fruit juice/pulp

and adjust the pH to 3.3-3.6 using citric acid. A pH meter may be necessary to establish the recipe but afterwards the ingredients may simply be weighed out.

For marmalade, or jams which contain fruit pieces, it is necessary to soak the peel or fruit for 2-3 days in a concentrated (60%) sugar solution. This causes the peel/fruit to achieve the same density as the preserve and, as a result, it is evenly distributed through the jar and does not float to the surface.

Boiling:

Pour the batch into a stainless steel boiling pan and heat as quickly as possible with constant stirring to prevent the product burning onto the pan. It is important to use stainless steel to prevent the acids in the preserve reacting with the pan and causing off-flavours. The mixture is boiled until the sugar content reaches 68%. This is most conveniently measured using a hand-held refractometer or a sugar thermometer (68% sugar corresponds to a jam temperature of 129°C). The correct sugar content is critical for proper gel formation; repeated checks with a refractometer or thermometer are needed to make sure that:

- the sugar level reaches 68% (otherwise mold will grow on the product or a gel will not form)
- that 68% sugar is not exceeded by a large amount (otherwise the jam will crystallize)
- the sugar concentration increases rapidly at the end of boiling and particular care is needed

Filling and packaging:

In many countries, there are laws concerning the weight of food sold in a package and accurate filling to the correct weight is therefore important. The weight, cleanliness of the container and appearance of the product after filling should be routinely checked. In particular it is important to avoid getting preserve around the rim of the jar as this may prevent a vacuum forming, and will look unsightly and attract insects.

The preserve should be hot-filled into suitable containers which are then sealed with a lid. Temperature of filling is important, too hot and steam will condense on the lid and drop down onto the surface of the preserve, this will dilute the sugar on the surface and allow mold growth. If the temperature is too low the preserve will thicken and be difficult to pour and a partial vacuum will not form in the jar. Ideally the temperature should be 82-85°C.

The packaging is likely to be one of the main costs involved in production. Ideally glass jars should be used with new metal lids. Metal cans are also suitable but very expensive. Cheaper alternatives include plastic (PVC) bottles or plastic (polythene) sachets. However, these cannot be filled with hot jam as they will soften or melt. Technical advice should be sought if these packs are being considered. It is possible to use paper, polythene, or cloth tied with an elastic band or cotton, to cover jam jars. The appearance of the product is however, less professional and there is a risk of contamination by insects. This is not recommended unless metal lids are impossible to obtain.

Finally, the jars are held upright and the gel is formed during cooling. This can be done by standing the jars on shelves, or more quickly using a low cost water cooler. A partial vacuum should form between the surface of the jam and the lid when the product cools. This can be seen by a slight depression in the lid. If a vacuum does not form it means that the jar is leaking or the temperature of filling is too low.

Equipment required: Mesh sieves, strainers, Jelly bags, Aluminum or enameled pan (for sugar syrup), Accurate 2 Kg scales (e.g. + or − 10 g), pH meter or pH paper (optional), spoons, jugs, knives, plastic buckets etc. Juice extractors, stainless steel boiling pan, hand held refractometer or sugar thermometer, gas bottles and burner, jar filler and capper, jar cooler (optional).

CHAPTER 4

PROCESSING AND PRESERVATION OF VEGETABLES AND THEIR PRODUCTS

INTRODUCTION

Foods normally spoil on storage. Any operation that prevents or retards its natural change is regarded as preservation.

Principles of preservation

Microorganisms are the single most important factor responsible for spoilage of food products. By a suitable method of preservation their activity is retarded or stopped and thus shelf life is extended. The different methods of preventing microbial spoilage are based on two principles.

a) Killing the organism present in a food product and preventing contamination by sealing it in a container.

b) Suppressing microbial growth by use of low temperature or chemical or by removing water.

Methods of fruit and vegetable preservation

1. Physical Methods:

    a) **Heat Processing**: Bacteria, yeast, molds and enzymes can all be destroyed by heat. Heating to appropriate temperature for a particular time in suitable containers is one of the methods of preservation. The process of sealing food in containers such as cans or bottles and preserving them by application of heat till the microorganisms are killed is called canning.

    b) **Removal of Heat**: Refrigeration and freezing are other techniques of preservation.

    c) **Drying and Dehydration**: Fruits and vegetables contain 70-95% water. Microorganisms can grow and multiply in the presence of moisture. By removing water, bacteria, yeasts and molds are unable to grow. This prevents decomposition and arrests microbial action completely. Sun drying and solar tunnel drying are the methods commonly used for drying fruits and vegetables. In dehydration, by hot air drying suitable temperature is maintained for drying on industrial scale.

2. **Chemical Methods**: Preservation by chemicals which retard, hinder or prevent microbial spoilage. Natural preservatives are sugar, salt, acid while as chemical preservatives are sulphur dioxide, benzoic acid etc. To act as a preservative, sugar must be used in large quantities. Spoilage bacteria will not develop in sugar solutions of 40 to 50 per cent, but certain yeasts and molds are able to develop in much higher concentrations. A high proportion of sugar is therefore, essential for growth of microorganisms. Preservation in brine is largely used for vegetables required later in manufacture of pickles. In vinegar, the active ingredient is acetic acid which is used in sufficient concentrations to check the development of destructive organisms.

    a) **Sulphur Dioxide**: Sulphur dioxide is used in preservation of fruit pulps, juices and concentrates. The concentration varies from 350-2000 mg/kg product. Potassium metabisulphite (KMS) is commercially available form of sulphur dioxide.

b) **Benzoic Acid:** It is available in market in the formulation of sodium benzoate. It is used for preservation of fruit juices, pulps and tomato products. The concentration varies from 0.06 to 0.10 per cent.

Biological Methods:

Lactic acid fermentation plays an important role in the preservation of vegetables in the form of pickles. Lactic acid bacteria converts' sugar to lactic acid which has a preserving effect on vegetables.

CANNING AND BOTTLING OF VEGETABLES IN BRINE

Select mature, sound and tender vegetables; wash them free of impurities with fresh water. Remove the inedible and damaged-portions.

Preparation: Prepare and grade the vegetables according to maturity and size as best as possible. Wrap the prepared vegetable in a piece of muslin cloth and dip in boiling water for 2-5 minutes. The time of blanching depends upon the type of vegetable and its maturity.

Blanching: Blanching is done to inactivate the enzymes, to drive out the air from tissues etc. Remove the muslin bag from the boiling water and dip it in cold water immediately. The water for blanching should be changed when it begins to froth. Usually 2% brine is used. It is prepared by dissolving 2 g of common salt in about 98 mL of water and filtering.

Filling and brining: Fill the blanched vegetable into tin cans or glass jars. Arrange the vegetable pieces in such a way that the container holds the given weight of vegetable. For A 23 size cans the fill-in weight should be 540-570 g. For butter cans >P 340-400 g. For pint glass jars >, 280-310 g. Fill the interspace in the can with 2 per cent hot and clear brine leaving a head space of 0.6-1 cm. Small quantities of sugar and citric acid are added to the brine while canning peas and cauliflower.

Exhausting: Place the filled cans in a boiling water bath with a false bottom (a frame work of wood or wire or a thick pad of cloth placed at its bottom); continue heating for 6-10 minutes till the centre of the can records a temperature of 75-82°C. The time of exhaust will vary with the nature of the pack.

Sealing: Immediately after exhausting, seal the cans with a can sealer. In the case of glass jars, the lids must have a rubber ring between the mouth of the jar and the top to achieve air-tight pack.

Processing: Vegetables in the sealed containers are processed processed i.e. heated at 116°C corresponding to 0.7 kg/sqcm pressure in a pressure cooker for 30-60 minutes. The processing time depends upon the size of the container, the kind of vegetable, its stage of maturity and the density of the pack. For every 150 m rise in altitude, the boiling point of water decreases by 0.6°C and hence a corresponding increase of 2 minutes for every 300 m rise in altitude should be effected in the processing time given in the schedule.

Cooling and storage: Place the cans immediately under running cold water and let the glass jars cool in the air. Store the product in a cool, dry place.

CANNING OF CURRIED VEGETABLES

Preparation: Prepare the vegetable as for use in the home kitchen, e. g., shell the peas; remove hard leaves of cauliflower and cut the flower head into flowerets of suitable size; remove the outer hard leaves of cabbage and cut into shreds of 0.6 - 1 cm thickness; peel the root vegetables and cut them into slices of suitable thickness, etc. Grade them according to size, quality, where practicable. Prepare the gravy as follows;

For one dozen A 24 cans of curried vegetables, weigh the following ingredients:

a. Mustard (whole) 20 g

b. Coriander (powder) 20 g

c. Red *chillies* (powder) 15-20 g

d. Caraway seed 20 g

e. Turmeric (powder) 40-50 g

f. Common salt (powder) 90 g

g. Vegetable fat (hydrogenated oil) 400 g

h. Water Sufficient to make up gravy for 12 cans

Heat the fat in a pan and fry mustard in it till the seeds crack. Add the other spices and continue frying for a few more minutes. Add the required quantity of water, stir thoroughly and bring the entire mass to boil.

Filling: Fill the prepared raw vegetables in A 23 size cans. Add hot gravy in proportions given below for some typical packs.

Combination of vegetables: Potatoes~cauliflower, potatoes cauliflower-tomatoes, potatoes-tomatoes, potatoes-peas, potatoes-peas-cauliflower

For other combinations, almost similar proportions maybe used. If, however, a product with fewer amounts of liquid portion is required, the proportion of vegetables may be increased and that of gravy decreased.

Exhausting, sealing, processing, cooling and storage: Except that the cooking time is 60-75 minutes depending upon the size, and type of containers, kind and maturity of vegetables, quantity of fat in the final product, etc. the other steps are exactly the same as in Canning and Bottling of Vegetables.

Drying of vegetables: The main step in the dehydration of vegetables is: preparation (washing, peeling and slicing), blanching, sulphiting and drying. Blanching may be carried out either in boiling water or in steam, but in the case of green vegetables like green peas, split field beans, spinach and fenugreek leaves, blanching in boiling water containing 0.1 per cent magnesium oxide, 0.1 per cent sodium carbonate and 0.5 per cent potassium metabisulphite (KMS) is recommended to ensure maximum retention of the natural green colour in the dehydrated vegetable. In other cases, sulphiting may be carried out immediately after blanching by steeping in half the weight of KMS solution of appropriate (0.10-0.25 per cent) concentration.

While beet-root is not sulphited, onions and garlic are neither blanched nor sulphited. For drying, generally about 5-10 kg of material are loaded per sq. m. of the area in trays.

Packing and storage: The dried vegetables should be put into confectionery tins and sealed air- tight with solder or wax, depending upon the duration of storage.

SWEET TURNIP

PICKLE

There are three types of turnips - red, white and yellow. Select tender, sound and evenly matured turnips free from fibres. Chop off the leaves. Wash the turnips thoroughly in fresh water and remove their thick outer peel. Cut them into 0.6 - 1.25 cm thick and round slices.

Recipe: Turnip slices 10 kg, red chilies 250 g, black pepper 125 g, mustard 500 g, spices (caraway and cinnamon) 62 g, dried dates (choara) 250 g, tamarind 250 g, ginger (fresh, chopped) 250 g, onion (fresh, chopped) 1 kg, garlic (fresh, chopped) 125 g, salt (powdered) 1 kg, vinegar (1 bottle) 800 mL, *gur* (jaggery) 625 g, rape seed oil (*sarson* oil) 1 litre.

Procedure: Fry the chopped ginger, onion and garlic in small quantity of oil. Prepare a thick extract of the tamarind pulp. Cut the dried dates into small pieces. Mix all the ingredients except gur and rape seed oil with the turnip slices. Fill the mass in a glass or glazed container. Place it in the sun for 3-4 days till the characteristic pickle flavour is developed. Add clarified thick syrup of jaggery to the pickle and leave in the sun for another 2 - 3 days. Add rape seed oil (previously heated and cooled) and keep the pickle in the sun for another 3-4 days till the slices have become sufficiently soft and palatable and the flavour and aroma of the pickle sufficiently enriched.

GREEN CHILLI

A. GREEN CHILLY AND LIME PICKLE

Select sound, fully mature and juicy limes having deep yellow skin and green chilies of good size. If the limes are greenish in colour, allow them to develop deep yellow colour before use. Wash the limes and chilled thoroughly in running cold water. Remove the stalks of green chilies without injuring their caps. Cut the limes into halves or quarters depending upon their size. For every kg of limes and green chilies used, weigh out about 250 g of powdered salt and transfer to a clean sterilized wide mouthed glass or glazed jar. Squeeze out some juice from limes over the salt. Transfer the partially squeezed limes to the salt. Make deep longitudinal slits in the green chilies and add them to the same jar. Stir the mass well to effect thorough mixing. If the juice of the limes is insufficient to cover the mass, add cool, boiled water acidified with citric acid (4-5 per cent) so that the limes and the chilies are covered by the liquid. Keep the mass in the sun for a week with occasional shaking, for softening the material. This is indicated by the skin of the lemons turning light brown and the green colour of the chilies also turning to brown. This pickle at this stage is ready for use. Keep off moisture from the pickle to prevent it from getting moldy and spoiled.

B. CANNED GREEN CHILLIES

Select, tender green chilies and snip off their stalks with a knife taking care to retain the calyx. Blanch the chilies for 2-3 minutes in boiling water. Rinse them in cold water and fill into sterilized cans; cover with hot 2 per cent brine (171 g chillies and 157 g brine per 1 lb jam size can). Add 50 mg citric acid to every 100 g of product. Exhaust the cans for 6-7minutes in hot water at 85 - 88°C. Seal and process for 30minutes in boiling water. Cool quickly and store in a cool and dry place.

CARROT

A. SPICED CARROT JUICE

Select sound, well matured and sweet carrots of the deep pink variety. Wash them thoroughly in water by scrubbing with hands. Remove the hairy parts. Chop off the two ends; scrape the carrot to remove the inedible portions completely. Grate the prepared material on a grater.

Recipe:

Grated carrots	:	10 Kg
Common salt	:	325 g
Mustard (powdered)	:	30 g
High quality vinegar	:	625 g
Red chillies (powdered)	:	3 teaspoons

Mix the ingredients thoroughly and pack the mass in earthen ware pitchers to the brim. Cover them with lids. Apply paraffin wax all-round the lids leaving a small opening on one side to let off the gases produced inside the pitchers; store in a cool, dry place for about a month, Press out the juice by hand wearing rubber gloves and strain through coarse muslin cloth. Allow the strained juice to stand in closed and deep containers for a couple of days. When the sediment separates from the clear juice, decant it and filter through fine muslin cloth. Add sodium benzoate (300 ppm) and mix thoroughly. Fill the clear juice into bottles previously sterilized in boiling water for 10-15 minutes. Cork tightly or crown seal the bottles. Store in a cool, dry place. Before use, dilute the beverage 2 - 3 times with water.

B. CARROT PRESERVE

Select fresh, sound and tender carrots of uniform size and colour. Wash them thoroughly in water by scrubbing with hands to remove the mud and other foreign materials sticking on them. Remove the hairy parts, chop off the two ends and scrape the skin with stainless steel' knives. Cut the peeled carrots into uniform slices of 5-8 cm long. Prick the slices with stainless steel forks thoroughly. Place the pricked carrots in sufficient boiling water to cover them and cook gently until they become just soft. Too much of cooking will spoil the shape and texture; little of cooking will result in slow penetration of sugar, dark colour and toughness. Drain off the water and spread the pieces on a clean white cloth for removing some moisture.

Prepare sugar syrup by dissolving 2 cups (8 oz. cup) of sugar in 3 cups of water, heat to boil and filter through thick muslin cloth. The quantity of syrup required should be equal to 24 times by weight of the quantity of the prepared carrots. Add the pieces to the boiling syrup and continue heating. When the syrup volume is reduced to half of the original, add one teaspoonful (5-6 g) of citric acid for every 1 kg of carrots; continue heating till the temperature reaches 106°C at sea level, or the syrup becomes thick enough to give 2-3 threads when drawn between the two fingers. Keep the mass for 48 hours. Boil again and fill into clean dry glass or glazed containers. Close the containers and store.

GINGER

Ginger (*Zingiber oficinale*) is the underground stem or the rhizome grown in South India. Ginger is used as flavoring for confectionery, ginger beer, chutneys and other culinary preparations.

GINGER CANDY

Select tender, fibreless and large rootlets or fingers, and wash them in cold water. Peel their skins with a sharp stainless steel knife and cut them into pieces of desired shape and size. Soften the ginger by: (i) boiling it in 0.5 per cent solution of citric acid (quantity of the solution being sufficient to cover the ginger) in a covered aluminum, stainless steel or heavily tinned brass or copper vessel for a period of six hours or (ii) by cooking in 0.5 per cent solution of citric acid at 0.7 kg per sq. cm. steam pressure in pressure cooker for one hour. Citric acid is used to bleach or whiten the colour during softening. Remove the ginger and wash it well with cold water. When sufficiently cooled, prick the softened pieces with stainless steel, or wooden prickers, Wash again. The ginger pieces are now ready for impregnation with sugar syrup.

Prepare 30° Brix syrup by dissolving 3 parts of sugar in 7 parts of water. Boil and filter the syrup through thick muslin cloth. Use about 1 kg of syrup for 1 kg of the prepared ginger. Boil the prepared ginger in the 30°Brix syrup for 15minutes and allow it to stand overnight, taking care that the ginger is completely covered by the syrup. After about 24 hours, drain off the syrup and increase its concentration to about 35°Brix by adding more sugar and heating, if necessary. Boil the ginger with the syrup for 15 - 20 minutes and keep it again overnight taking care to have the ginger fully covered by the syrup. Repeat this every day till the concentration of the syrup is about 60°Brix. At this stage, add a small quantity (about 0.1 per cent of the total weight of the syrup) of citric acid or tartaric acid, or 5 per cent by weight of invert sugar or corn syrup, carry out the process of absorption by increasing the strength of the syrup by 5°Brix each day, till it reaches 75°Brix. The preserve is then ready to be packed.

Set aside the prepared preserve for 2-3 months for thorough penetration of sugar into the ginger. Then boil it for about 5 minutes. While still hot, drain off the syrup and roll the pieces on finely ground sugar. Place the pieces on a wooden tray and dry them in shade or at 50°C in a drier till they are no longer sticky. The syrup left over after candying can be used again for preparing more ginger preserve, or used as a syrup for flavoring aerated waters.

GINGER PICKLE

Wash the ginger with plenty of water using a brush to remove all soil residues. Peel the ginger and again wash in clean water. Cut the ginger vertically in 2 mm thick slices. Blanch the ginger in boiling water for 15 minutes and drain off the hot water. Repeat this process four times. Mix with 1% salt, pack the mixture in a jar and pour lemon juice (or vinegar) until the product is totally immersed. Keep it for one week to become sour and the ginger turns to the color pink. The flow sheet is given as;

Delivery

↓

Sorting

↓

Washing

↓

Peeling

↓

Washing

↓

Cutting (2 mm slices)

↓

Blanching (15 minutes at 90°C)

↓

Mixing (with 1% salt, pour lemon juice or vinegar until product is totally immersed)

↓

Packing

↓

Storage (for one week until it becomes sour and colour of ginger turns into pink)

GINGER PRESERVE (SALTY)

Wash the ginger with plenty of water using a brush to remove all soil residues. Peel the ginger and again wash in clean water. Cut the ginger vertically in 2 mm thick slices. Blanch the ginger in boiling water for 15 minute and drain off the hot water. Repeat this process seven times. Mix with 10% salt, dissolve and then put on the low, medium heat. Stir every now and then until salt crystals appear on the ginger. Move from the heat and cool it before packing. The flow sheet is given below,

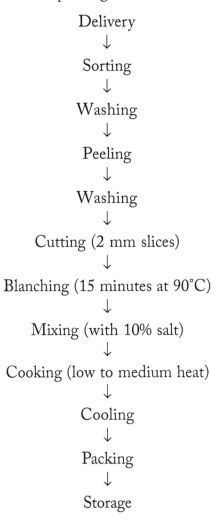

Delivery
↓
Sorting
↓
Washing
↓
Peeling
↓
Washing
↓
Cutting (2 mm slices)
↓
Blanching (15 minutes at 90°C)
↓
Mixing (with 10% salt)
↓
Cooking (low to medium heat)
↓
Cooling
↓
Packing
↓
Storage

GINGER PRESERVE (SWEET AND SOUR)

Wash the ginger with plenty of water using a brush to remove all soil residues. Peel the ginger and again wash in clean water. Cut the ginger vertically in 2 mm thick slices. Blanch the ginger in boiling water for 15 minutes and drain off the hot water. Repeat this process seven times. Mix with 60% sugar, dissolve and then put on the low medium heat. Pour two teaspoons of lemon juice for every one kilogram of ginger and stir every now and then until sugar crystals appear on the ginger. Move from the heat and cool before packing.

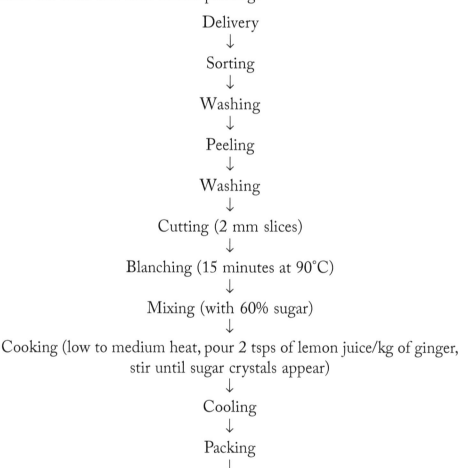

Delivery
↓
Sorting
↓
Washing
↓
Peeling
↓
Washing
↓
Cutting (2 mm slices)
↓
Blanching (15 minutes at 90°C)
↓
Mixing (with 60% sugar)
↓
Cooking (low to medium heat, pour 2 tsps of lemon juice/kg of ginger, stir until sugar crystals appear)
↓
Cooling
↓
Packing
↓
Storage

GINGER PRESERVE (SWEET)

Wash the ginger with plenty of water using a brush to remove all soil residues. Peel the ginger and again wash in clean water. Cut the ginger vertically in 2 mm thick slices. Keep in a lye solution for 30 minutes, then wash in clean water and keep in a saturated salt solution for 24 hours. Every eight hours, drain off the water and replace with fresh and clean water over two days. Blanch the ginger in boiling water for 15 minutes and drain off the hot water. Repeat this process seven times. Sugar (60% by weight) is dissolved in clean water, on a low, medium heat. When all the sugar dissolves, filter through a thin cloth and mix the ginger with the purified sugar solution, keeping for one day. The following day, when it has released its moisture, the ginger is taken from the sugar solution. Boil the sugar solution until the volume becomes half and move it from the heat and cool it. When the solution becomes cool, keep the ginger in it for one day. Repeat this procedure three times. After three days, the mixture is put on the low medium heat, stir continuously until the solution becomes thick and small sugar crystals appear on the ginger. Move from the heat and cool before packing.

Delivery → Sorting → Washing → Peeling → Washing → Cutting → Soaking → Blanching → Mixing → Screening → Boiling the sugar solution → Mixing → Cooking → Cooling → Packing → Storage

CANNING AND BOTTLING OF PROCESSED PEAS

Pea (*Pisum sativum*) is extensively cultivated as a field and garden crop in India. The crop is very popular because of its sweet pods which are used in fresh condition in various preparations. Select dried peas of uniform size and quality. Reject all grains attacked by insects. Soak sound grains in 3 times the quantity of water for 15-18 hours. Reject the hard grains which have not absorbed any water. Wash the soaked peas thoroughly to remove extraneous matter. Wrap the soaked grains in a piece of muslin cloth and then dip in boiling water for 5 minutes so that the grains are soft enough to be pressed by hand. Remove the muslin bag from the boiling water when it begins to froth.

Fill the blanched peas into sterilized cans or glass jars in this way: 450 g for A 24 size can; 240 g for pint glass jars; and 220 g for butter size can. Fill the inter space with hot and clear solution containing 2 per cent common salt, 4 percent sugar and a little mint flavour and requisite quantity of edible green colour. Leave a head space of 0.6-1 cm. Exhaust the cans until the centre of the can attains a temperature of 82°C (time taken is 7-10 min) and seal. For pasteurization, the sealed cans are processed at lip C (0.7 kg steam pressure) in a pressure cooker for 40 minutes in the case of A2 and cans and glass jars and 30 minutes for butter cans. For every 300 m rise in altitude, increase the processing time by two minutes. Cool the cans in running water and jars in air.

CANNING OF BAKED BEANS AND PULSES IN TOMATO SAUCE

Several pulses like Bengal gram (*Cicer arietinum*), blackgram (*Phaseolus mungo*), cowpea or lobia (*Vigna unguiculata*), horsegram (*Dolichos biflorus*) and French bean (*Phuseolus vulgaris*) widely used in India can be canned in tomato sauce.

Preparation of tomato sauce:

Wash fresh tomatoes of proper ripeness and good colour and cut them into pieces. Heat for 10 minutes in their own juice in a stainless steel pan. Pass the heated material through a screw extractor to separate the skin and seeds. Use the pulp obtained for the sauce.

Recipe:

Tomato pulp (6 per cent solids), Cardamom, pepper and jeera in equal quantities, Cinnamon, Cloves (only stalks), Mace (not ground)s alt, Onions (chopped), Garlic (ground), Sugar, White vinegar, Red chilli powder.

Procedure

Add the quantity of sugar given in the recipe. Place the spices in a muslin bag and immerse it into the pulp. Cook the pulp till it is reduced to half the original volume. Remove the muslin bag and squeeze it into the pulp. Add vinegar, salt and remaining sugar. Beat the mass for a few minutes so that the volume of the finished product is half of the original pulp. Fill the finished sauce into clean, sterilized A 24 size plain cans, seal them and process for 45 minutes in boiling water and store for subsequent use. For use with beans, dilute the sauce with half its weight of water and add corn starch (1 percent by weight of the diluted sauce) to improve the consistency of canned product.

Pretreatment of beans:

Soak the beans in water for 6 - 8 hours. Change water twice to prevent souring. Blanch the beans in 1 per cent sodium bicarbonate solution for 15 - 20 minutes. Wash in cold water to remove the residual bicarbonate taste. Bake the blanched beans in steam at 1.05 h./sq. cm. pressure in a pressure cooker. The beans turn brownish in colour and become soft. Some beans also break up and impart a melting texture to the canned product.

Canning of beans in sauce:

Fill 190 g of baked beans directly into plain 1 lb jam size cans and cover with 170 g of hot tomato sauce. Exhaust for 7 minutes at 85 - 88°C (temperature at the centre of the can will be 74 - 77°C), seal and process for 75 minutes at 0.7 kg/sq.cm pressure. Cool in running cold water and store for further use. The canned product keeps well in storage for about 5 months.

Canning of pulses in sauce:

Soak the pulses in water for 4-6 hours and blanch for five minutes in boiling water. After blanching, bake French beans and Eobia for 10 minutes at 0.7 kg/sq.cm pressure. Fill the blanched or baked material into 1 lb jam size cans and cover with hot tomato sauce (190 g pulse and 170 gsauce per cari) exhaust the cans for 7 minutes at 85-88°C. Seal the cans and process for 75 minutes at 0.7 kg/sq cm pressure. Cool in running water and store.

PETHA

CANDY

Select sound, fully ripe Petha with tough texture. Cut it into slices and remove the seeds and inner soft pulp. Remove the outer hard surface with a stainless steel knife. Cut the peeled slices into cubes or pieces of any desired shape and size. Prick the cubes or pieces of *petha* with a stainless steelfork or bamboo pricker. Keep them completely immersed in fresh lime water (prepared from 60 g of quick lime to 1 Litre of water, vigorously stirring allowing it to settle and finally decanting and filtering) for 3-4 hours, depending upon the softness of the petha. The softer the petha, the longer is the period for which it is to be left. Drain off the lime water and wash the fruit pieces thoroughly in fresh water.

Soften the petha pieces by placing them in boiling water for about 15 - 30 minutes so as to make them take up sugar from the syrup. Drain off the water and spread the pieces on a clean white sheet for removing some moisture. Prepare sugar syrup by dissolving 2 parts of sugar in 3 parts of water, heating to boil, remove the scum and finally filter it through thick cloth to get a clear syrup. The quantity of syrup prepared should be equal to three times the quantity of the prepared fruit. Add the fruit pieces to the boiling syrup and continue heating till the temperature reaches 107°C (at sea level) or the syrup becomes thick enough to give 2 to 3 threads when drawn between the two fingers. Allow the whole, mass to stand overnight; drain off the syrup through a stainless steel sieve. Add essence if desired and store it in a clean dry glass or glazed container.

BAMBOO

Bamboos are found in the forests of Assam, Bengal, Bihar, Madhya Pradesh, Orissa, Tamil Nadu, Kerala and Karnataka. Bamboo shoots (tender) are used as an article of food by the poorer classes of people especially during famine. For making edible products, use stunted or mist–shaped shoots which are not likely to produce bamboos of good quality select tender bamboo shoots 45 - 60 cm long. Remove the sheaths or outer covering leaves with a sharp knife. Cut the tender portions into rings or pieces of suitable size. Thinly scrape off any green portion on the pieces (leafy portions towards the growing tip may be used for making chutney). Boil the rings or pieces of shoots in water 2 - 3 times for half an hour each time, to remove the bitterness. Change the water every time. Prick the boiled pieces with a stainless steel needle or fork. The bamboo pieces prepared in this way be used for candies, chutneys and for canning.

BAMBOO CANDY

Prepare syrup of 30°Brix by mixing 3 parts of sugar with 7 parts of water and straining. Use about 1.75 kg of syrup per kg of bamboo pieces. Cover the pricked rings or pieces of shoots with the syrup and boil for a few minutes. Allow them to stand in the syrup for 24 hours. If available determine the concentration of sugar in the syrup by means of a Brix hydrometer. The Brix will be slightly less than 30° due to the absorption of sugar by bamboo shoots. Drain the syrup and increase its concentration by about 10°Brix by adding more sugar. Bring the syrup to boil and pour it back on the shoots. Repeat this every day until the concentration of the syrup reaches about 60°Brix. At this stage, add a small quantity of citric or tartaric acid (about 0.1 per cent of the total weight of syrup). Increase the strength of the syrup by 5°Brix each day till it reaches 75°Brix. Keep the product in the syrup for about a week. Boil the shoots along with the syrup for about 5 minutes. While still hot, drain the syrup and roll the pieces in finely ground sugar. Add any flavour, if needed. Place the pieces on a wooden tray and dry them in shade. Pack in dry containers and seal air-tight. Store in a cool and dry place. If desired, individual pieces may be wrapped in cellophane paper.

BAMBOO CHUTNEY (SWEET)

The bamboo pieces prepared for candy making may be mixed with the tender leafy portion and utilized for chutney. After removing the bitterness, mince them in a meat mincer or chop them finely with a sharp knife.

Recipe: Minced or chopped tender shoots, sugar, salt, cardamom, cinnamon and cumin in equal quantity, red chilies, finely ground onions, finely chopped garlic, finely chopped vinegar (good quality).

Method: Add sugar, salt and a little water to the minced bamboo shoots and warm in an aluminum pan till sugar and salt have gone into solution. Tie the spices loosely in a thick muslin bag and leave the bag in the pan. Continue cooking till the material has softened and acquired the consistency of a jam. At this stage, remove the bag from the pan and press its contents. Add vinegar, and boil again to the desired consistency. Cool the product a little; pack it into clean, dry wide mouthed jars and seal them air-tight. Store in a cool, dry place.

Canning of bamboo in syrup:

Prepare 40° Brix syrup containing 1 per cent citric acid by mixing 2 kg of sugar, 5 g of citric acid and 3 kg of water and filtering the mixture. Bring the syrup to boil and pour into the cans filled with bamboo pieces, leaving 0.6 cm head space. Place the cans in boiling water so that the space at the top of the can is 2.5-3.75 cm above the level of water. Heat the cans for 5 - 6 minutes or until the temperature in the centre of the can is about 93°C. Remove the cans from the bath and seal them immediately with a can sealer. Process the sealed cans in a pressure cooker at 0.7 kg pressure or at 116°C as follows: 1 lb jam size can 40 minutes, 1 lb butter size can 50 minutes. Mix 23 cans 50 minutes. After processing, cool the cans in running cold water, wipe them dry and store them in a cool place. Dry the place.

Canning of bamboo in brine

Fill the prepared pieces into plain cans, keeping about 1 cm head space. Prepare 2.0 per cent common salt solution by using 20 g of retied table salt per liter of water. Bring the solution to boil, strain it through cloth, and pour hot into the cans, leaving 0.6 cm head space. The rest of the stages are the same that are followed for the canning of bamboos in syrup.

Canning of bamboo in curried vegetables:

Prepare the bamboo pieces in the usual way. Prepare the other vegetables also in the usual way prescribed for them. For preparing the liquid medium for the vegetables use ingredients in the proportion given below for 48 cans:

Mustard (whole) 85 g, Coriander (powder) 85 g, Red chillies (powder) 55-85 g, Caraway seeds 85 g, Water as necessary, Turmeric (powder) 170-200 g, Common salt 370 g, Vegetable fat (hydrogenated oil) 1.6 kg. Heat the required quantity of fat in a pan till a few grains of mustard, when added to it, produce a crackling sound. Then, add the entire quantity of mustard grains and fry for a few minutes. Add the other spices and fry. Add the requisite quantity of water, stir thoroughly and bring to boil.

Fill plain A2 cans with the prepared material in any of the combinations mentioned below, leaving 0.3 cm head space. Add the hot liquid medium.

Place the filled cans immediately in a boiling water bath for 5-6 minutes, or until the temperature at the centre of the can reaches 82 - 88°C. Remove the cans from the bath, seal them immediately with a can sealer.

Process the sealed cans in a pressure cooker at 0.7 kg pressure or at 116°C for 70-75 minutes depending on the kind and maturity of the vegetables and the quantity of fat in the final product. After messing, cool the cans in running cold water, wipe them and store in a cool, dry place.

GARLIC

Dried Garlic

Garlic cloves are peeled and washed with a lot of water and then strained. The cloves are sliced, spread on a sieve tray and then dried in the dryer. Dried garlic slices are packed in polyethylene bag. The dried garlic is sold in various forms such as sliced granules or powdered and used as a spice mixture for sauce, soup mixtures, salad dressing, meat products and garlic salt. Flow diagram for the product is as:

Delivery → Sorting → Peeling → Washing → Cutting → Drying → Packing → Storage

LEMON

Lemon Syrup

Ripe fruits are selected. After washing, peel and cut into halves. With the manual extractor the juice is separated from the skin and pips. Dissolve sugar (60% by weight) in clean water, boil for five minutes, stirring all the time; and then filter the impurities from the sugar solution. Add the lemon juice and boil for two minutes. Move from the heat and fill the bottle while hot. Cooling, cleaning and labeling are the last operations before storage.

Delivery → Sorting → Washing → Peeling → Cutting → Extraction → Filtration → Mixing → Cooking → Bottling → Capping → Sterilization → Cooling → Washing → Labelling → Storage

Lemon Preserve (Sour)

Blanch lemons for fifteen minutes and repeat this three times. Soak in boiled water for one day. The following day, drain off the water and dry in the sun dryer.

Delivery → Sorting → Blanching → Soaking → Drying → Packing → Storage

Lemon Preserve (Sweet)

Soak the lemons in the boiled water and drain off after 12 hours, repeat twice a day for three days. The following day, blanch and cut into quarters, soak in 5% salt solution for five days and drain. 100% by wt of sugar are melted, filtered, and then mixed with the lemon quarters for one night. The next day, lemon quarters are removed from the sugar solution and then the sugar solution is boiled on low, medium heat until the volume becomes half. The lemon is mixed with the syrup and placed on a low fire until all the water boils out. Move from the heat and cool down before packing.

Delivery → Soaking → Blanching → Mixing → Packing → Storage

Lemon Pickle

Blanch the lemon for fifteen minutes and repeat this three times. Soak in the boiled water for one day. The following day, drain off the water and dry in the sun dryer. Roast the mustard seeds and grind. Cook the edible oil and cool down. The dried lemon is mixed with the sliced ginger, small piece of garlic, chilli and spices. The product is cooled down before packing.

Delivery → Sorting → Blanching → Soaking → Drying → Cooking → Mixing → Packing → Storage

ONION

Dried Onion

Onion varieties with pungent flavor are most appreciated; both colored and white onions may be used. After removing the tops, roots and outer skin, onions are washed carefully then cut at a right angle to the core of the onions. Blanching is not practiced, as the onion loses its flavor. The use of preservatives is not necessary: after cutting, the slices are spread evenly on the trays of a dryer. The onions are dried when the ratio of prepared raw material to drier product is about 9:1 (moisture content 5%). The dried product may be ground to a powder, which tends to make clumps. The dryer used for onions must be reserved especially for onions.

Delivery → Sorting → Peeling → Washing → Cutting → Drying → Packing → Storage.

POTATO

Potato Chips

Peel off potato skins and put potatoes in cold water containing 0.05% KMS to prevent browning reaction. Cut into thin slices and keep in clean water. Put them individually separated on the mosquito screen in the sun. Fry in hot oil until slices are crispy. Set aside to cool. Pack in plastic bags. Seal each bag properly to ensure shelf life.

Delivery → Peeling → Soaking → Cutting → Drying → Frying → Cooling → Packing → Storage.

Dried Potato Slices

After sorting potatoes, wash, peel, and cut them into 2 mm thick slices. For preservation purposes and in order to keep the color, blanching is carried out. Keeping on the bamboo sieve, the potatoes are dipped in boiling water containing 5 grams salt per litre of water, and 3 grams potassium metabisulphite per litre of water. Then drain and spread them on the trays of the dryer. The potato slices are dried when the prepared raw material/dried product ratio is about 10:1.

Delivery → Peeling → Soaking → Cutting → Drying → Packing → Storage

MIXED VEGETABLE PICKLE

Recipe:

A) **Materials required :**

1)	Prepared knoll-khol	1 kg
2)	Prepared carrot	1 kg
3)	Prepared Radish	1 kg
4)	Prepared Cauliflower	1 kg
5)	Prepared Nadroo	1 kg

B) **Spices needed:**

1)	Red Chilli powder	100 gm
2)	Clove	5 gm
3)	Zeera	10 gm
4)	Cardamom	5 gm
5)	Mustard Seeds	100 gm
6)	Turmeric	40 gm
7)	Ajwain	

8) Onion (100 g), garlic (25 g) and ginger (fresh 100 g) past

9) Salt as per taste.

C) **Edible oil: 800 mL**

D) **Acetic acid: 20 mL or vinegar: 500 mL**

Method:

1) Cut the vegetable into desired shape and wash them thoroughly.

2) Tie material in a piece of muslin cloth and blanch by immersing in hot boiling water containing 1 tbs turmeric to provide colouration.

3) Take out muslin cloth from water; allow draining water and drying on clean tray in sun for 2- 3 hours to remove excess water.

4) Clean mustard seeds, ajwain and zeera and roast on hot tawa on moderate heat.

5) Fry garlic, onion and ginger paste in small quantity of oil, add spices except the vegetable pieces turn slightly soft allow them to cool.

6) Add mustard seeds. When the spices turn brown, add the prepared vegetables. When the vegetable pieces turn slightly soft allow them to cool.

7) Add 20 mL Glacial acetic acid or 500 mL Vinegar mix thoroughly and lace it in the sun for 3-4 more days.

8) Fill the pickle in pre-sterilized glass jars tightly with spoon and fill the gaps with oil. Pour small quantity of oil on top to provide oil seal which avoids contamination.

9) Keep in sun for 2-3 days so that proper aroma, texture and taste is developed.

LEMON OR LIME BARLEY WATER

Recipe:

Lime or lemon Juice	:	4 kg
Barley extract	:	1 kg (from 15 g of barley flour)
Sugar	:	2. 5 kg
Preservative (KMS)	:	2.75 g

Method:

1. Wash, cut fruit into halves and extract juice in lime juice squeezer. In case of lemons, extract juice with hands or use juice extractor.

2. Strain the juice through coarse muslin cloth to remove seeds and coarse pulp.

3. Mix about 15 g of barely flour with some water and make it into a thin paste. Continue to add water and stir till the whole quantity of water (1 kg) is used. Heat gently to disperse the barely flour thoroughly. Cool and strain through muslin cloth.

4. Fill the product into clean bottles. Screw tight the lids and seal them by dipping the top end in molted paraffin was.

5. Store in cool and dry place till use.

SPICED CARROT JUICE

Recipe:

Prepared Carrot (Grated)	:	5 kg
Salt	:	150 g
Mustard (Powdered)	:	15 g
Glacial acetic acid	:	12 g
Or		
Vinegar	:	300 g

Method:

1. Select tender and sweet carrots of the deep pink variety.

2. Wash, scrape to remove hairy parts and cut off tips and bottom ends.

3. Grate the prepared carrots finely on an aluminium or s/s graters.

4. Mix all the ingredients and pack the mixture in earthen ware pots (pitchers) to the brim. Cover with lids.

5. Seal with paraffin wax along the lids leaving only a small opening on one side to let off gases produced inside the pots.

6. Store for one month in a cool, dry place. Then press out the juice and strain through the muslin cloth.

7. Allow the juice to stand in closed and deep containers for two days. Decant the clear juice and filter through muslin cloth.

8. Dissolve sodium benzoate (300 mg/kg of the beverage) in a small quantity of water and mix thoroughly with the product.

9. Fill juice in clean, sterilized bottles leaving one inch head space.

10. Cork or screw tight or crown seal the bottles. Dip the top lid in melted paraffin wax to seal the bottles when using corks or old screw type lids.

11. Store in a cool and dry place.

12. Dilute the beverage 2-3 times in water before use.

GINGER TONIC

Recipe:

Ginger Juice	:	250 g
Sugar	:	650 g
Water	:	4.1 litre
Citric acid	:	17 g

Method:

1. Scrape ginger rhizomes, wash and grate into fine pieces, Extract juice with the help of muslin cloth.

2. Mix water and sugar. Warm slightly and add citric acid. Strain through muslin cloth, cool syrup.

3. Add ginger juice to the syrup and mix thoroughly.

4. Leave the product for maturation for about two days at room temperature. Keeping the product covered with muslin cloth.

5. Decant clear ginger tonic, pasteurize and fill into sterilized juice bottles leaving 1 inch headspace.

6. Crown cork bottles immediately and sterilize bottles in boiling water for 20 minutes.

7. Remove the bottles from boiling water, cool and store them in a cool dry place.

8. Chill in refrigerator before use.

CHAPTER 5

PROCESSING AND PRESERVATION OF CEREAL AND CEREAL PRODUCTS

WHEAT: Varieties and Characteristics

The three principal types of wheat used in modern food production are *Triticum vulgare* (or *aestivum*), *T. durum*, and *T. compactum*. *T. vulgare* provides the bulk of the wheat used to produce flour for bread making and for cakes and biscuits (cookies). It can be grown under a wide range of climatic conditions and soils. Although the yield varies with climate and other factors, it is cultivated from the southernmost regions of America almost to the Arctic and at elevations from sea level to over 10,000 feet. *T. durum*, longer and narrower in shape than *T. vulgare*, is mainly ground into semolina (purified middlings) instead of flour. Durum semolina is generally the best type for the production of pasta foods. *T. compactum* is more suitable for confectionery and biscuits than for other purposes.

The wheat grain, the raw material of flour production and the seed planted to produce new plants, consists of three major portions: (1) the embryo or germ (including its sheaf, the scutellum) that produces the new plant, (2) the starchy endosperm, which serves as food for the germinating seed and forms the raw material of flour manufacture, and (3) various covering layers protecting the grain. Although proportions vary, other cereal grains follow the same general pattern. Average wheat grain composition is approximately 85 percent endosperm, 13 percent husk, and 2 percent embryo.

The outer layers and internal structures of a kernel of wheat: *Encyclopaedia Britannica, Inc.* Characteristic variations of the different types of wheat are important agricultural considerations. Hard wheats include the strong wheats of Canada (Manitoba) and the similar hard red spring (HRS) wheats of the United States. They yield excellent bread-making flour because of their high quantity of protein (approximately 12–15 percent), mainly in the form of gluten. Soft wheats, the major wheats grown in the United Kingdom, most of Europe, and Australia, result in flour producing less attractive bread than that achieved from strong wheats. The loaves are generally smaller, and the crumb has a less pleasing structure. Soft wheats, however, possess excellent characteristics for the production of flour used in cake and biscuit manufacture.

Wheat intermediate in character include the hard red winter (HRW) wheat of the central United States and wheat from Argentina. There are important differences between spring and winter varieties. Spring wheats, planted in the early spring, grow quickly and are normally harvested in late summer or early autumn. Winter wheats are planted in the autumn and harvested in late spring or early summer. Both spring and winter wheats are grown in different regions of the United States and Russia. Winter varieties can be grown only where the winters are sufficiently mild. Where winters are severe, as in Canada, spring types are usually cultivated, and the preferred varieties mature early, allowing harvesting before frost.

In baking and confectionery, the terms strong and weak indicate flour from hard and soft wheats, respectively. The term strength is used to describe the type of flour, strong flours being preferred for bread manufacture and weak flours for cakes and biscuits. Strong flours are high in protein content, and their gluten has a pleasing elasticity; weak flours are low in protein, and their weak, flowy gluten produces a soft, flowy dough.

The protein content and major food uses of certain varieties of wheat: *Encyclopaedia Britannica, Inc.*Wheat breeders regularly produce new varieties, not only to combat disease but also to satisfy changing market demands. Many varieties of wheat do not retain their popularity, and often those popular in one decade are replaced in the next. New varieties of barley have also been developed, but there have been few varieties of rice.

WHEAT FLOUR

The milling of wheat into flour for the production of bread, cakes, biscuits, and other edible products is a huge industry. Cereal grains are complex, consisting of many distinctive parts. The objective of milling is separation of the floury edible endosperm from the various branny outer coverings and elimination of the germ, or embryo. Because wheats vary in chemical composition, flour composition also varies. Although some important changes have occurred in flour milling, basic milling procedure during the past 100 years has employed the gradual reduction process as described below;

Milling:

In modern milling considerable attention is given to preliminary screening and cleaning of the wheat or blend of wheat to exclude foreign seed and other impurities. The wheat is dampened and washed if it is too dry for subsequent efficient grinding, or if it is too damp it is gently dried to avoid damaging the physical state of the protein present, mainly in the form of the elastic substance gluten.

The first step in grinding for the gradual reduction process is performed between steel cylinders, with grooved surfaces, working at differential speeds. The wheat is directed between the first "break," and set of rolls, and is partially torn open. There is little actual grinding at this stage. The "chop," the resulting product leaving the rolls, is sieved, and three main separations are made: some of the endosperm, reduced to flour called "first break flour"; a fair amount of the coarse nodules of floury substances from the endosperm, called semolina; and relatively large pieces of the grain with much of the endosperm still adhering to the branny outsides. These largish portions of the wheat are fed to the second break roll. The broad objective of this gradual reduction process is the release, by means of the various sets of break rolls, of inner endosperm of the grain, in the form of semolina, in amounts sufficient that the various semolinas from four or five break rolls can be separated by suitable sieving and the branny impurities can be removed by air purifiers and other devices. The cleaned semolinas are reduced to fine flour by grinding between smooth steel rolls, called reduction rolls. The flour produced in the reduction rolls is then sieved out. There are usually four or five more reduction rolls and some "scratch" rolls to scrape the last particles of flour from branny stocks. Since the various sieving and purification processes free more and more endosperm in the form of flour, flour is obtained from a whole series of processing operations. The flour is sieved out after each reduction roll, but no attempt is made to reduce to flour all the semolina going to a particular reduction roll. Some of the endosperm remains in the form of finer semolina and is again fed to another reduction roll. Each reduction roll tends to reduce more of the semolina to flour and to flatten bran particles and thus facilitate the sieving out of the branny fractions. The sieving plant generally employs machines called plan-sifters, and the air purifiers also produce a whole series of floury stocks. Modern flour processing consists of a complicated series of rolls, sieves, and purifiers. Approximately 72 percent of the grain finally enters the flour sack.

The sacked flour may consist of 20 or more streams of flour of various states of purity and freedom from branny specks. By selection of the various flour streams it is possible to make flour of various grades. Improvements in milling techniques, use of newer types of grinding machinery in the milling system, speeding up of rolls, and improved skills have all resulted in flour produced by employing the fundamentals of the gradual reduction process but with simplified and shorter milling systems. Much less roll surface is now required than was needed as recently as the 1940s.

The purest flour, selected from the purest flour streams released in the mill, is often called patent flour. It has very low mineral (or ash) content and is remarkably free from traces of branny specks and other impurities. The bulk of the approximately 72 percent released is suited to most bread-making purposes, but special varieties are needed for some confectionery purposes. These varieties may have to be especially fine for production of specialized cakes, called high-ratio cakes that are especially light and have good keeping qualities. In many countries the flour for bread production is submitted to chemical treatments to improve the baking quality. In modern processing, regrinding of the flour and subsequent separation into divisions by air treatment has enabled the processors to manufacture flour of varying protein content from any one wheat or grist of wheats.

Composition and grade:

Flour consists of moisture, proteins (mainly in gluten form), a small proportion of fat or lipids, carbohydrates (mainly starch, with a small amount of sugar), a trace of fibre, mineral matter (higher amounts in whole meal), and various vitamins. Composition varies among the types of flour, semolinas, middlings, and bran.

Protein content:

For bread making it is usually advantageous to have the highest protein content possible (depending on the nature of the wheat used), but for most other baked products, such as cookies (sweet biscuits) and cakes, high protein content is rarely required. Gluten can easily be washed out of flour by allowing a dough made of the flour and water to stand in water a short time, followed by careful washing of the dough in a gentle stream of water, removing the starch and leaving the gluten. For good bread-making characteristics, the gluten should be semi-elastic, not too stiff and unyielding but not soft and flowy, although a flowy quality is required for biscuit manufacture. The gluten, always containing a small amount of adhering starch, is essentially hydrated protein. With careful drying it will retain its elasticity when again mixed with water and can be used to increase the protein content of specialized high-protein breads. Sometimes locally grown wheat, often low in protein, may be the only type available for flour for bread making. This situation exists in parts of France, Australia, and South Africa. The use of modern procedures and adjustment of baking techniques, however, allow production of satisfactory bread. In the United Kingdom, millers prefer a blend of wheat, much of it imported, but modern baking procedures have allowed incorporation of a larger proportion of the weak English wheat than was previously feasible.

Treatment of flour:

Use of "improvers," or oxidizing substances, enhances the baking quality of flour, allowing production of better and larger loaves. Relatively small amounts are required, generally a few parts per million. Although such improvers and the bleaching agents used to rectify excessive yellowness in flour are permitted in most countries, the processes are not universal. Improvers include bromates, chlorine dioxide (in gaseous form), and azodicarbonamide. The most popular bleacher used is benzoyl peroxide.

Grade:

The grade of flour is based on freedom from branny particles. Chemical testing methods are employed to check general quality and particularly grade and purity. Since the ash (mineral content) of the pure branny coverings of the wheat grain is much greater than that of the pure endosperm, considerable emphasis is placed on use of the ash test to determine grade. Bakers will generally pay higher prices for pure flour of low ash content, as the flour is brighter and lighter in colour. Darker flours may have ash content of 0.7 to 0.8 percent or higher. A widely employed modern method for testing flour colour is based on the reflectance of light from the flour in paste form. This method requires less than a minute; the indirect ash test requires approximately one to two hours.

BAKING

Baking is common and important for food, both from an economic and cultural point of view. A person who prepares baked goods as a profession is called a baker. All types of food can be baked, but some require special care and protection from direct heat. Various techniques have been developed to provide this protection. In addition to bread, baking is used to prepare cakes, pastries, pies, tarts, quiches, cookies, scones, crackers, pretzels, and more. These popular items are known collectively as "baked goods," and are often sold at a bakery, which is a store that carries only baked goods, or at markets, grocery stores, farmers markets or through other venues.

Meat, including cured meats, such as ham can also be baked, but baking is usually reserved for meat loaf, smaller cuts of whole meats, or whole meats that contain stuffing or coating such as bread crumbs or buttermilk batter. Some foods are surrounded with moisture during baking by placing a small amount of liquid (such as water or broth) in the bottom of a closed pan, and letting it steam up around the food. Roasting is a term synonymous with baking, but traditionally denotes the cooking of whole animals or major cuts through exposure to dry heat; for instance, one bakes chicken parts but roasts the whole bird. One can bake pork or lamb chops but roasts the whole loin or leg. There are many exceptions to this rule of the two terms. Baking and roasting otherwise involve the same range of cooking times and temperatures. Another form of baking is the method known as *encroute* (French for "in a pastry crust"), which protects the food from direct heat and seals the natural juices inside. Meat, poultry, game, fish or vegetables can be prepared by baking *encroûte*. Well-known examples include Beef Wellington, where the beef is encased in pastry before baking; pate encroute, where the terrine is encased in pastry before baking; and the Vietnamese variant, a meat-filled pastry called pâté chaud. The *encroute* method also allows meat to be baked by burying it in the embers of a fire – a favorite method of cooking venison. Salt can also be used to make a protective crust that is not eaten. Another method of protecting food from the heat while it is baking is to cook it *en papillote* (French for "in parchment"). In this method, the food is covered by baking paper (or aluminum foil) to protect it while it is being baked. The cooked parcel of food is sometimes served unopened, allowing diners to discover the contents for themselves which adds an element of surprise.

Eggs can also be used in baking to produce savoury or sweet dishes. In combination with dairy products especially cheese, they are often prepared as a dessert. For example, although a baked custard can be made using starch (in the form of flour, corn flour, arrowroot, or potato flour), the flavor of the dish is much more delicate if eggs are used as the thickening agent. Baked custards, such as crème caramel, are among the items that need protection from an oven's direct heat, and the *bain-marie* method serves this purpose. The cooking container is half submerged in water in another, larger one, so that the heat in the oven is more gently applied during the baking process. Baking a successful soufflé requires that the baking process be carefully controlled. The oven temperature must be absolutely even and the oven space not shared with another dish. These factors, along with the theatrical effect of an air-filled dessert, have given this baked food a reputation for being a culinary achievement. Similarly, a good baking technique (and a good oven) is also needed to create a baked Alaska because of the difficulty of baking hot meringue and cold ice cream at the same time.

Baking can also be used to prepare other foods such as pizzas, baked potatoes, baked apples, baked beans, some casseroles and pasta dishes such as *lasagne*.

Equipment:

Baking needs an enclosed space for heating – typically in an oven. The fuel can be supplied by wood, coal, gas, or electricity. Adding and removing items from an oven may be done by hand with an oven mitt or by a peel, a long handled tool specifically used for that purpose. Many commercial ovens are equipped with two heating elements: one for baking, using convection and thermal conduction to heat the food, and one for broiling or grilling, heating mainly by radiation. Another piece of equipment still used for baking is the Dutch oven. "Also called a bake kettle, bastable, bread oven, fire pan, bake oven kail pot, tin kitchen, roasting kitchen, *doufeu* (French: "gentle fire") or *feude compagne* (French: "country oven"). It originally replaced the cooking jack as the latest fireside cooking technology," combining "the convenience of pot-oven and hangover oven. Asian cultures have adopted steam baskets to produce the effect of baking while reducing the amount of fat needed.

BAKERY AND BAKED GOODS

These include categories like bars, breads (bagels, buns, rolls, biscuits and loaf breads), cookies, desserts (cakes, cheesecakes and pies), muffins, pizza, snack cakes, sweet goods (doughnuts, Danish, sweet rolls, cinnamon rolls and coffee cake) and tortillas.

CAKE

APPLE CAKE

Ingredients: Butter-175 g, plus extra for greasing, eggs–3, self raising flour-350 g, ground cinnamon-2 tsp, sugar-175 g, apples-3 medium, dates, halved, stoned and finely chopped- 100 g, blanched hazelnuts, roughly chopped-100 g.

Method: Heat the oven to 180°C. Lightly butter a deep 20 cm loose-based or spring form round cake in, then line the base with baking parchment. Melt the butter by heating it and then cool it for 5 minutes. Crack the eggs into the butter and beat well. Mix the flour with the cinnamon and the sugar. Core and cut two apples (unpeeled) into bite-size chunks. Mix the apples into the flour along with the dates and half of the chopped hazelnuts. Pour the egg and butter mixture into the flour mixture and gently stir together. Spoon into the tin, smooth the top. Thinly slice the remaining apple (unpeeled) into circles, discard the pips, and then arrange over the top of the cake. Sprinkle over the remaining Cakes & Desserts 8 hazelnuts. Bake for 50 mins-1 hr until the cake is cooked and risen. Check if it is done by pushing a skewer into the centre – it should come out clean. Remove the cake from the tin, peel off the paper and transfer the cake to a wire rack. Cool completely and keep for 3 days.

BLACK FOREST CAKE

Ingredients: Refined flour-60 g, cocoa powder-60 g, baking powder-¼ tsp, Sodium bicarbonate- a pinch, sugar-120 g, eggs-4, melted butter-120 g, vanilla essence-2 drops. For decoration: whipped cream-200 g, grated chocolate-15 g, walnuts or broken cashew nuts or cherries- 50 g.

Method: Line and grease two 175 mm diameter cake tins. Sift flour, cocoa, sodium bicarbonate and baking powder together twice. Beat eggs and sugar over hot water till thick like custard sauce. Add vanilla essence. Fold in lightly the sieved flour and melted butter, adding a little water to get pouring consistency. Put the mixture in the tins and bake at 400° F for about 20 minutes. Cool and sandwich together with the fresh cream and put a thick layer of cream on top.

COOKIES

Ingredients: Maida: 2½ cups, butter: 1 cup, sugar: 1 cup, egg: 1, milk: 2 table spoons, vanilla essence: 1 teaspoon, salt: a pinch

Method: Preheat oven to 190°C. Lightly coat 2 cookie sheets with vegetable oil. Sift flour, baking powder and salt together. Beat egg yolks in a mixer bowl until pale and thick. In a clean mixer bowl, with clean beaters, beat egg whites to soft peaks. Beat in sugar 1 table spoon at a time, until stiff and glossy. Gently fold egg yolks into egg whites. Fold in dry ingredients and milk until just blended. Drop by level tablespoonfuls 2 inches apart onto prepared cookie sheets. Bake for 10 minutes or until golden. Carefully transfer to wire racks to cool.

TYPES OF COOKIES

1. Drop/short Cookies: These are made from relatively soft dough that is dropped by spoonful on to the baking sheet. During baking, the mounds of dough spread and flatten. Chocolate chip cookies, oatmeal cookies and rock cookies are popular examples of drop cookies.

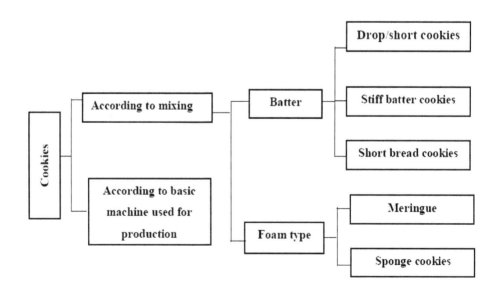

Recipe:

Ingredients		Quantity
Maida	:	2½ cups
Butter	:	1 cup
Sugar	:	1 cup
Egg	:	1
Milk	:	2 table spoons
Vanilla essence	:	1 teaspoon
Salt	:	a pinch

Method:

1. Preheat oven to 190°C.
2. Lightly coat 2 cookie sheets with vegetable oil.
3. Sift flour, baking powder and salt together.
4. Beat egg yolks in a mixer bowl until pale and thick.
5. In a clean mixer bowl, with clean beaters beat egg whites to soft peaks.
6. Beat in sugar 1 table spoon at a time, until stiff and glossy.
7. Gently fold egg yolks into egg whites.
8. Fold in dry ingredients and milk until just blended.
9. Drop by level table spoonful's 2 inches apart onto prepared cookie sheets.

10. Bake for 10 minutes or until golden.

11. Carefully transfer to wire racks to cool.

2. Stiff Batter Cookies: These are prepared from stiff dough. It is made stiffer by refrigerating before cutting and baking. Then rolled into cylinders which are sliced into round cookies before slicing. Pinwheel cookies are the best example.

3. Meringue Cookies: These are light, airy, sweet, and crisp because whipped egg whites and sugar are the base ingredients. It is a gluten-free sweet cookie, without flour.

4. Sponge Cookies: These are light and airy like the meringue cookies but whole egg is used instead of only egg whites.

Stiff Batter Cookies:

Meringue Cookies:

Sponge Cookies

BREAD

Ingredients: 450 g of strong white flour, 25 g of butter, salt (a pinch), sugar (1 tsp), yeast (2 tsp-fast acting)

Methods

Step 1: Sieve 450 g of strong white flour into a large bowl then add 25 g of butter and rub it in with your fingertips. Now add a good pinch of salt, 1 tsp of sugar and 2 tsp fast action yeast and stir everything together.

Step 2: Make a well in the centre then pour in 300 mL pint of lukewarm water, and stir together with a wooden spoon. Then use your hands to bring the dough together. It will be very sticky at this point so add a little extra flour as needed.

Step 3: Now turn the dough out onto a floured surface and begin to knead for about 10 minutes; put the heel of your hand on the dough and push away from you then turn the dough over and pull it back to you. You will soon develop a rhythm. The dough should become smooth and silky.

Step 4: Place the dough in a clean, lightly oiled large bowl. Cover with cling film and put it somewhere warm for it to rise for about 1 hour. It should double in size.

Step 5: Now knock back the dough, this means to knead the dough again for a couple of minutes. This is so the yeast is distributed evenly in the dough. Shape the dough to fit a 2 lb loaf tin then grease and lightly flour the tin and put the dough seam side down into it. Loosely cover with lightly oiled cling film and leave it to prove somewhere warm for 30 minutes or until it has doubled in size.

You can slash the loaf if you wish and dust with flour. Put it in a pre-heated oven at gas mark 8/23°C (22°C in a fan oven) on the middle shelf and bake for about 30-35 minutes. Remove bread from the tin and leave to cool on a wire rack.

TIPS

- If kneading seems like too much hard work add the mixture at step 2 to a food mixer and mix with the dough hook for about 5-10 minutes until it wraps its self around the hook.

- To test whether the bread is cooked, it should sound hollow when tapped. If you prefer the bottom of the bread to be crisp, once removed from the tin put it back in the oven for a further 5 minutes.

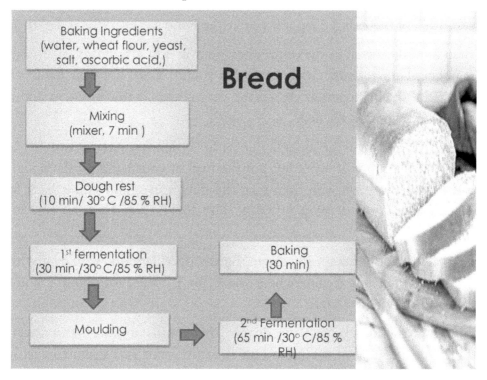

PIZZA

¼ ounce active dry yeast, 1 tsp sugar, 1-1/4 cups warm water (110-115°C), ¼ cup canola oil, 1 tsp salt, 3-1/2 to 4 cups all-purpose flour, ½ ground beef/mutton, 1 small onion, 1 can (15 ounces) tomato sauce, 3 tsps dried oregano, 1 tsp dried basil, 1 medium green pepper, diced, 2 cups shredded part-skim mozzarella cheese

Methods. In large bowl, dissolve yeast and sugar in water; let stand for 5 minutes. Add oil and salt. Stir in flour, 1 cup at a time, until a soft dough forms. Turn onto floured surface; knead until smooth and elastic, 2-3 minutes. Place in a greased bowl, turning once to grease the top. Cover and let rise in a warm place until doubled, about 45 minutes. Meanwhile, cook beef and onion over medium heat until no longer pink; drain. Punch down dough; divide in half. Press each into a greased 12-in. pizza pan. Combine the tomato sauce, oregano and basil; spread over each crust. Top with beef mixture, green pepper and cheese. Bake at 205°C for 25-30 minutes or until crust is lightly browned.

CHAPTER - 6

PROCESSING AND PRESERVATION OF DAIRY AND DAIRY BASED PRODUCTS

KHOA

Khoa is prepared by different methods depending on the location and quantity of milk available for conversion.

PEDA

Peda is a sweet prepared from *pindi* variety of *khoa* by the addition of sugar. Since *peda* contains sugar and lower moisture content it has a better keeping quality than *khoa*. *Peda* have religious importance as they are offered as ⁻*Prasad* during worship of God in the temples. *Peda* is also offered to guests at the time of ceremonial celebration like marriages etc. Some region specific varieties are popular in different regions of the country. *Doodh peda* is the common variety and is popular all over India. *Peda* is characterized as a circular slightly flattened ball with low moisture content and white to creamy white in colour and smooth texture

Flow diagram for Preparation of *peda* from *khoa*

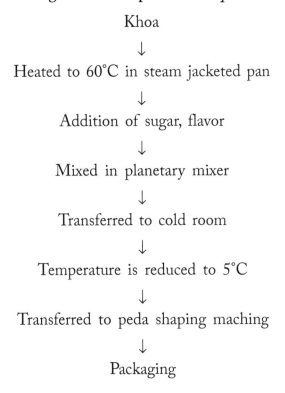

Khoa

↓

Heated to 60°C in steam jacketed pan

↓

Addition of sugar, flavor

↓

Mixed in planetary mixer

↓

Transferred to cold room

↓

Temperature is reduced to 5°C

↓

Transferred to peda shaping maching

↓

Packaging

Method for Preparation of Doodh Peda

Traditional method: About 5 litres of buffalo milk is taken in an open pan and heated on a brisk non smoky, fire. When the milk reaches a rabri stage, about 400 - 450 gm sugar is added and stirring and scraping continued until a pasty consistency is obtained. During the last stages of heating, the paste is worked up into a smooth mass. The heating is then stopped; the paste is spread on the walls of the pan for cooling. Then the product is shaped in to either flattened circular balls or rectangular shapes.

Preparation of *peda* from *khoa*: *Pindi khoa* is broken into bits and heated to 90°C. Powdered sugar @ 30% on the basis of *khoa* is added and heating is continued with rigorous working to obtain a smooth pasty consistency. Then the product is cooled and shaped on molds.

BURFI

Methods of Production: Buffalo milk is preferred for making *burfi*. Milk used for *burfi* should not have objectionable flavours and titratable acidity should not be more than 0.16 percent. Milk is filtered before use to remove visible objectionable foreign matter. Standardized buffalo milk with 6% fat and 9% SNF in quantities of 4-5 lit per batch is taken in a double jacketed stainless steel kettle and heated. Milk is boiled continuously with constant stirring and scraping so as to avoid burning of solids on the surface of the kettle. When a semisolid consistency is attained, heating is discontinued. Powdered sugar @ 30% on the basis of *khoa* is added and blended thoroughly into *khoa* with the help of a flattened wooden ladle. When a homogeneous mass with desirable flow characteristics is achieved, the blend is transferred to greased trays. The product is allowed to set for minimum of 4 hours. Then *burfi* is cut into desirable shapes and sizes with a knife and packed *burfi* is stored at room temperature.

**Preparation of *burfi* from pre-made *khoa:* *Burfi* can also be prepared from pre made *khoa* obtained from market. Hot sugar syrup is prepared from 300 gm sugar by adding minimum quantity of water. Hot sugar syrup is added to 1 kg *khoa* and heated to 80°C. The mixture is kneaded properly and when desirable flow characteristics are attained, it is poured in trays and allowed to set at room temperature.

Manufacture of *burfi* from concentrated milk: Acceptable quality *burfi* can be prepared from concentrated milk. Concentrated milk with 35% total solids is taken and heated in a kettle. When a semisolid consistency is reached, ground sugar @ 30 percent on *khoa* weight basis is added and kneaded. When desirable flow characteristics are seen in the product, the mix is emptied into trays and set-aside.

Preparation of *burfi* from cream and skim milk powder: Cream with 30 per cent fat and skimmed milk powder are mixed in 1:2 ratio. The mixture is heated to 92°C and heating is continued till 75 per cent total solids are attained in the product. The heating is then stopped;

ground sugar is added at 50°C and kneaded. Then mixture is poured in trays for setting.

Flow diagram for the manufacture of *burfi* from standardized buffalo milk

Buffalo milk

Filtration/Clarification

Standardization (6% fat and 9% SNF)
Boiling in a kettle with stirring and scraping

Attainment of semi solid consistency

Stop heating

Addition of powdered sugar

Kneading with wooden laddle till desirable
flow charactristics are attained

Pouring in trays

Setting for 4 hours overnight at room temperature

Cut into desirable shapes and sizes

Packaging

↓

Storage at room temperature

Mechanized production of *burfi:* A mechanized process for commercial production of *burfi* was successfully developed by the NDDB. All the ingredients, like *khoa*, sugar, additives, such as cardamom, etc, are first heat processed to blend uniformly in a planetary mixer. The processed ingredients are then fed to a Rheon-shaping and forming machine. A die is placed at the end of the encrusting. Machine gives shape to the *burfi* emerging out from the machine in a continuous uninterrupted flow. *Burfi* is then packed.

KALAKAND

Preparation of *kalakand* from milk: Buffalo milk is preferred for *kalakand* manufacture. Slightly sour milk (up to 0.18% lactic acid) can be used for its preparation. Buffalo milk standardized to 6% fat and 9% SNF is taken in a pan and boiled. At the appearance of first boiling, 0.05% citric acid (on weight of milk) dissolved in small quantity of water is added to milk. There is no need to add citric acid in case of slightly acidic milk. The milk is boiled with continuous stirring and scraping. At pat formation stage, sugar @ 30% on expected yield of *khoa* or alternatively 7.5% on the basis of milk is added and stirring is continued. When desirable textural and body characteristics are achieved, mixture is removed from fire and poured in a tray, smeared with a thin layer of ghee for setting. The *kalakand* is cut into desirable shapes or alternatively served as such without any definite shape.

Preparation of kalakand from *khoa*: *Danedar khoa* is taken in a pan and heated. To this sugar @ 30% is added and heating is continued with stirring. When desirable flow characteristics are observed, *kalakand* is poured into a tray and set for minimum 4 hr.

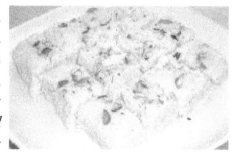

Flow diagram for the preparation of *kalakand*

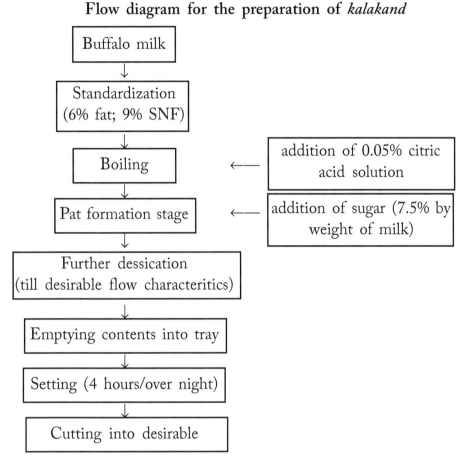

Chemical Composition: The gross composition of *kalakand* is nearer to the composition of *burfi*. Market samples carry higher amounts of moisture and lower percentage of fat than lab made samples. The chemical composition of *kalakand* chiefly depends on quality of milk, method of production and amount of sugar added.

Milk cake

Method of production: Whole buffalo milk is boiled in an open pan. When milk comes to first boil, 0.05% citric acid on milk basis is added as 1% solution. After addition of citric acid solution, small tiny granules tend to appear in milk. Heating is continued further with rigorous stirring.

When the mass shows signs of leaving the surface of pan, sugar @ 50 percent of expected *khoa* yield 12.5% on the basis of milk is added and the mix is heated with rigorous stirring. Controlled heating is required at this stage to obtain better quality product. When the product shows signs of dry appearance, the mix is poured in deep metal molds while hot. The milk cake is allowed to cool slowly at room temperature so that the central portion of the pat turns brown and caramelized flavour develops due to residual heat. After cooling, the product is cut into desirable sizes. At times, the vessel containing the hot mass is placed in a container with water so that the surface of milk cake cools fast giving white appearance outside and brown coloration in the central core.

Flow diagram for the preparation of Milk Cake

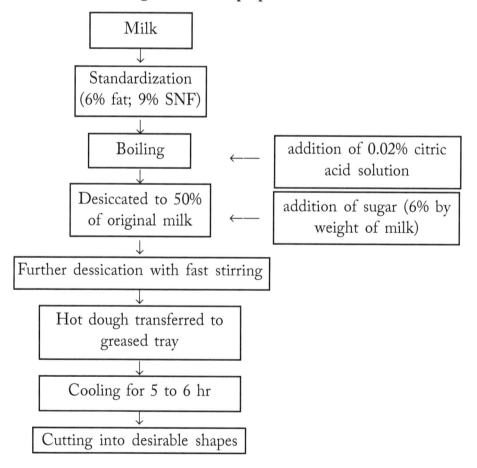

Packaging of milk cake: The pieces of milk cake are generally packed in vegetable parchment lined paper board boxes. Individual pieces are packed in LDPE pouches.

GULABJAMUN

Method of preparation: *Dhap* variety of *khoa, maida* and baking powder (750 g *khoa*, 250 g *maida* and 5 g baking powder) are blended to form homogenous and smooth dough. Small amount of water can be added in case of hard dough and if it does not roll into smooth balls. The mix should be prepared fresh every time. The balls are then deep fat fried at 140° C to golden brown colour and transferred into 60% sugar syrup maintained at 60 °C. It takes about 2 hours for the balls to completely absorb sugar syrup.

Mechanized production of *Gulabjamuns*: A mechanized semi-continuous system, developed by NDDB, for the manufacture of *Gulabjamun* from *khoa* employs meat ball forming machine and potato chip fryers.

Flow diagram for manufacture of *Gulabjamun* from *khoa* using meat ball forming machine

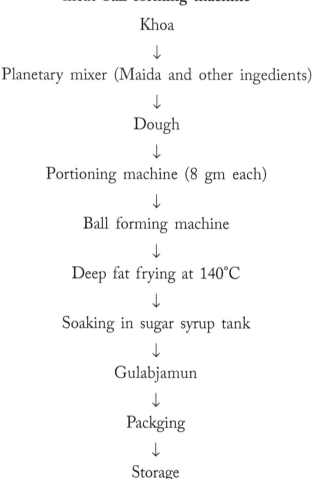

Khoa

↓

Planetary mixer (Maida and other ingedients)

↓

Dough

↓

Portioning machine (8 gm each)

↓

Ball forming machine

↓

Deep fat frying at 140°C

↓

Soaking in sugar syrup tank

↓

Gulabjamun

↓

Packging

↓

Storage

Preparation of *Gulabjamun* from instant *jamun* mix: The existing method of *Gulabjamun* preparation is suitable for cottage scale and cannot be adopted for large scale production. During summer months and festive seasons, *khoa* in required quantities may not be available to prepare *Gulabjamuns*. To meet *khoa* shortage and to produce *Gulabjamun* on a commercial scale, instant *Gulabjamun* mix was developed by Central Food Technology Research Institute, Mysore long back using spray dried skimmed milk powder, maida, vanaspati, citric acid, tartaric acid and baking soda. A complete *Gulabjamun* mix was also formulated in National Dairy Research Institute, Karnal based on roller dried skimmed milk.

RABRI

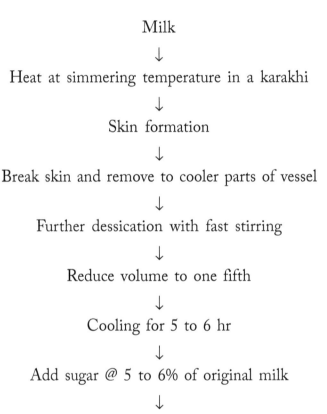

Milk
↓
Heat at simmering temperature in a karakhi
↓
Skin formation
↓
Break skin and remove to cooler parts of vessel
↓
Further dessication with fast stirring
↓
Reduce volume to one fifth
↓
Cooling for 5 to 6 hr
↓
Add sugar @ 5 to 6% of original milk
↓
Immerse layers of skin in the sweetened condensed milk
↓
Heat gently for a while

Traditional method: Whole buffalo milk is taken in a karahi and slowly heated at simmering temperature. The milk is not allowed to boil so that clotted cream may form on its surface. Slow evaporation of moisture takes place and the layer of resulting clotted cream is continuously removed with a fork and placed on the sides of *karahi*. When the volume of milk is considerably reduced, sugar @ 5-6% milk is added and stirred. The clotted cream collected on the sides of the pan is gently added back to the sweetened concentrated milk. The clotted cream is broken into small bits by wooden ladle before addition to milk. Whole mass then appears to be concentrated sweetened milk with clotted cream. Higher the amount of clotted cream better will be the quality of *rabri*.

Improved batch method: A small scale process was developed for the manufacture of *rabri*. The process includes heating of whole buffalo milk with 6% fat in a steam jacketed stainless steel kettle to 90-95°C and holding at this temperature for some time without any agitation. The *malai* (pellicle) formed on the surface of milk may be removed in a separate container after every 10 min. After collecting about 500 g *malai*, the temperature of milk is raised to boiling. Pre-determined quantity of sugar (@ 6% on milk basis) is added to the concentrated milk and mixed well. Subsequently, *malai* collected in a separate vessel is added back and mixed. The product is then cooled to < 10°C and served.

Large scale manufacture of rabri: Two types of concentrated milks were used as base materials for large scale production of *rabri*. In one process, buffalo milk with 2% fat was concentrated to 35% TS in a scraped surface heat exchanger. In another method buffalo skim milk was concentrated in a vaccum pan to contain 35% total solids. The concentrated milks were heated upto 90°C followed by addition of sugar @12% of the product. The mixture was then cooled to 70ºC and calculated amount of *malai* is added to it. The *malai* could be prepared from buffalo milk by simmering it in an open kettle. However *rabri* obtained by these two methods was liked slightly by judges indicating necessity for further improvement in the process.

BASUNDI

Buffalo milk is preferably used for *basundi* owing to its high total solids content. Traditionally it is prepared by heating whole milk in a pan over fire. Milk is thickened through evaporative heating with occasional scraping at the bottom. With progressive boiling, more and more thickening of milk occurs. The milk is concentrated to the consistency of the condensed milk and sugar (about 15-17% of concentrated milk) is added and stirred into the milk until it is fully dissolved. The pan is removed from the fire, allowed to cool and the flavouring material is added. Powdered cardamom (about 0.02% of concentrated milk) is added and mixed, along with saffron and borneol (edible camphor) (about 0.02% each of concentrated milk) and stirring continued till the desired consistency is achieved. The end-product has a pleasant caramel flavour and thick consistency. It is usually served chilled.

PAYASAM/KHEER

Method of Preparation: Buffalo milk is preferred for both *payasam* and *kheer* preparation. Standardized buffalo or cow milk with 4.5%-5.5% fat and 8.5 to 9% SNF is taken in a pan and boiled on a non-smoky fire. Good quality rice is taken, washed and added to milk @ 2.5%. Gentle boiling accompanied by thorough stirring cum scraping of the contents is undertaken. When the ratio of concentration of milk reaches 1: 1.8, sugar is added @ 5-7.5% of the milk taken. Further heating with stirring is continued until the rice is properly cooked and approximately when the concentration reaches 2 to 3 times. Powdered cardamom (@0.02%) is mixed as flavoring at the end of heating.

Dried kheer mix

Method of preparation: Fresh buffalo milk is standardized to 6.9 % fat and 9.5% SNF. The milk is preheated at 60°C for 15 min in the pre-heater of triple effect evaporator and then concentrated to 35% TS. Homogenisation of milk concentrate is done at a pressure of 183 bar in the first stage and 36 bar in the second stage. Then it is mixed with ground rice and sugar and slurry is heated to a temperature of 80 °C for gelatinizing of rice in as steam jacketed vessel. Using a fluidized bed dryer the slurry is dried and instantized. The product is packed in

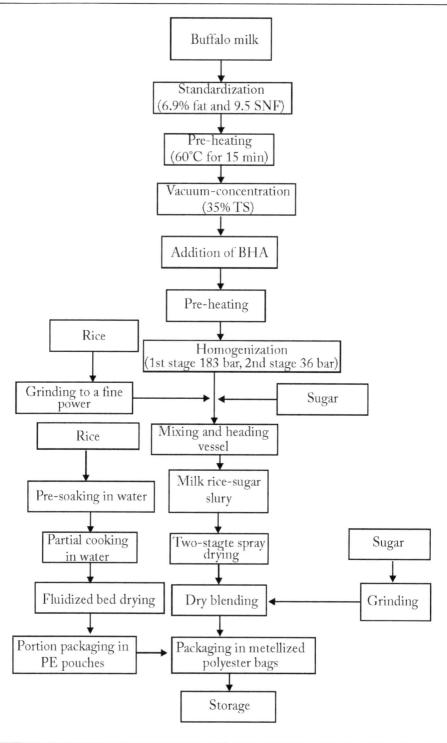

metalized polyester LDPE bags. Reconstitution of *kheer* mix involves rehydration of instant rice in boiling water for 10 min followed by dispersal of the powdered component into the rice water mixture. Then the product is garnished with dry fruits, flavors etc.

CHANNA

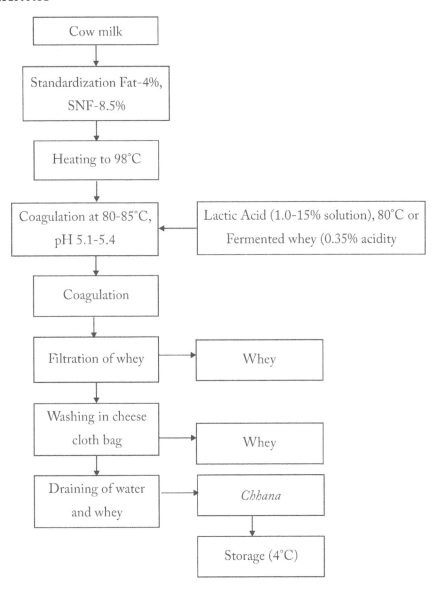

Traditional method of manufacture: *Chhana* has been prepared by boiling about 15-40 litre of cow milk in a steel pan. Acidic whey (previous day whey) added to boiling hot milk serve as coagulant with continuous stirring till the completion of coagulation. Contents poured over a muslin cloth held over another vessel. Whey is collected in a vessel. Muslin cloth containing curd mass washed with potable water by immersion process and allowed to drain for 30min to expel free whey.

Industrial production of *chhana* from cow milk: In industrial production, multi-purpose stainless steel vats or storage tanks are used for storage of milk, plate heat exchanger or steam jacketed kettle are used for heating of milk. Other process controls like temperatures of heating of milk, coagulation and coagulant are very precisely maintained as shown in flow diagram depicted below. SS strainers with cloth lining are used to filter the whey out of coagulum.

RASOGOLLA

Method of preparation: *Rasogolla* is prepared from soft, fresh cow milk *chhana*. Kneading of *chhana* to smooth paste by manually or using planetary mixer is first step in *Rasogolla* making. The smooth paste is portioned and rolled between palms to form balls of about 15mm diameter each weighing about 8-10 g in weight. Each ball should have smooth surface without visible cracks on surface. On an average, one kg *chhana* yields 90-100 *rasogolla* balls. These *rasogolla* balls are cooked in sugar syrup of approximately 50°Brix. Heating is regulated to maintain stability of the balls. Balls are cooked for 14 – 15 min. During cooking small amount of water is continuously added to maintain syrup concentration. This makes up for the loss of water due to evaporation. About 10% of sugar syrup should be replaced with fresh one each time to cook another batch. After cooking *rasogolla* balls are transferred to dilute sugar syrup at 60°C for texture and colour improvement. After 30 min stabilized balls are transferred to 60°Brix syrup for 1-2 hours, followed by final dipping in 50°Brix syrup.

Chemical composition of *rasogolla*

S. No.	Parameter	*Rasogolla* made from cow milk	*Rasogolla* made from buffalo milk
1	Moisture (%)	54.0 – 56.0	41.5 – 42.1
2	Total solids (%)	44.0 – 46.0	57.9 – 58.5
3	Protein (%)	5.0 – 5.2	5.3 – 5.8
4	Fat (%)	4.8 – 5.0	7.8 – 8.0
5	Carbohydrate (%)	33.6 – 35.1	43.6 – 44.1
6	Ash (%)	0.75	0.7
7	pH	6.50 - 6.70	6.67 - 6.79

Flow diagram for manufacture of *rasogolla*

Chhana

↓

Mixing & kneading into dough
in a planetary mixer

↓

Formed into balls (8-10 g)
In portioning machine

↓

Balls cooked in sugar syrup for 14-15 min

↓

Texture stabilization in dilute sugar syrup for 30 min

↓

Transferred to 60°Brix sugar syrup

↓

Finally transfer to 50° brix sugar syrup → Canning

↓

Cool to 10°C for
Refrigerated storage

Mechanized production of Rasogolla: Mechanical disc grinder has been designed for better kneading of *chhana*. Screw conveyor with kneading section and cutter provided at the exit simultaneously perform the kneading and portioning of *chhana* into lumps of about 10 g each. This lump is allowed to fall on a spinning disc, which has a stationary mounting above it. Rotation of lumps of *chhana* with stationary mounting forms it into the spherical balls. Recently rotating plate over a conveyor belt is also in use for ball formation. Now these balls can be made to fall directly into sugar syrup for cooking. Alternatively mechanized cooker can also be used for cooking; advantages are uniform weight and shape of *rasogolla* balls, large production, and uniform quality.

RASOMALAI

Method of preparation: *Chhana* is kneaded into smooth dough along with 1 to 4 % wheat flour. Dough is portioned and rolled into balls having a smooth texture without cracks. These balls are flattened to differentiate from *rasogolla* balls, Flattened balls are processed like *rasogolla* and subsequently stored in sweetened (5 – 6% sugar) milk thickened to one third of its volume. *Rabri* without the creamy layer can also be used instead of thickened milk to soak the cooked flattened balls. *Rasmalai* has limited shelf life of 3-5 days at refrigerated temperature.

Flow diagram for manufacture of *Rasmalai*

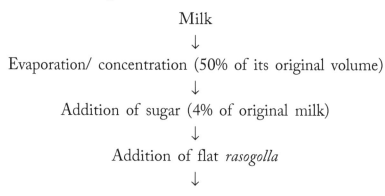

Milk

↓

Evaporation/ concentration (50% of its original volume)

↓

Addition of sugar (4% of original milk)

↓

Addition of flat *rasogolla*

↓

Heating (2-5min)

↓

Cooling/chilling

↓

Packaging

↓

Storage at < 5°C

PANEER

Traditional method of manufacture: Buffalo milk is boiled in a bigger iron vessel and a small portion of this is transferred to a smaller vessel. The coagulant (usually sour whey) is added to hot milk and stirred with a ladle till coagulation is completed. The contents of the vessel are emptied over a piece of coarse cloth to drain off whey. The whole process is repeated till all the milk is coagulated. The curd is collected after draining the whey and pressed to remove more whey. Finally, product is then dipped in chilled water.

Industrial method for *paneer* making: Buffalo milk is standardized to 4.5% fat and 8.5% SNF (standardize the buffalo milk to a fat: SNF ratio of 1:1.65). Milk is heated to 90°C without holding (or 82°C with 5 min holding) in a jacketed vat and cooled down to 70°C. Coagulation is done at about 70°C by slowly adding 1% citric acid solution(70°C) with constant stirring till a clean whey is separated at (pH 5.30 to 5.35) and coagulum is allowed to settle for 5 min and drained off the whey. The curd so obtained is filled into hoops lined with muslin or cheese cloth. Pressure is applied on top of the hoop at a rate of 0.5 to 1 kg/cm^2. The pressed blocks of *paneer* are removed from the hoops and immersed in pasteurized chilled water for 2-3 hr. The chilled *paneer* is then removed from water to drain out. Finally *paneer* blocks are wrapped in parchment paper / polyethylene bags and placed in cold room at about 5 to 10°C.

Flowchart depicting traditional method of *paneer* manufacturing

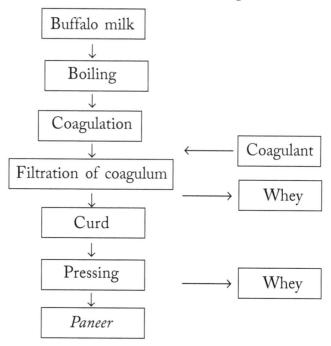

PANEER FROM COW MILK

Cow milk yields an inferior product in terms of body and texture. It is criticized to be too soft, weak and fragile and unsuitable for frying and cooking. Buffalo milk contains considerably higher level of casein and minerals particularly calcium and phosphorous, which tends to produce hard and rubbery body while cow milk produces soft and mellow characteristics. By replacing one third of buffalo milk with cow milk, a good quality *paneer* can be made. Buffalo milk *paneer* retains higher fat, protein and ash content and lactose as compared to cow milk *paneer*. To make *paneer* exclusively from cow milk, certain modifications in the conventional procedure have to be made. Addition of calcium chloride at the rate of 0.08 to 0.1% to milk helps in getting a compact, sliceable, firm and cohesive body and closely knit texture. A higher temperature of coagulation (85°-90°C) with coagulation of milk at pH 5.20 to 5.25 helps in producing good quality *paneer* from cow milk. However, at this pH of coagulation, moisture, yield and solids recovery are less.

Flowchart depicting industrial method of *paneer* manufacturing

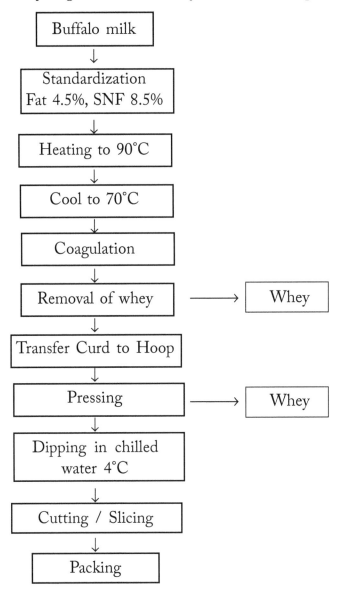

DAHI

Traditional method: In traditional method of dahi preparation, milk is heated intensively to boil for 5 to 10 min and then it is cooled to room temperature. Cooled milk is added with previous day's curd or buttermilk, stirred and allowed to set undisturbed usually for overnight. At halwai's shop milk is considerably concentrated before being inoculated with starter culture. So that the total solid content of milk gets increased, particularly increase in the protein content of milk. Concentration of milk results in custard like consistency of dahi and keeps the product from wheying off.

Method of preparation of Dahi

Receiving of milk

↓

Preheating (35 – 40°C)

↓

Filtration/Clarification

↓

Standardization – (Fat: 0–5%, SNF: 11–13)

↓

Preheating (60°C)

↓

Homogenization (175 kg/cm^2)

↓

Heat treatment (90° C/10 min)

↓

Cooling to 30°C

↓

Addition of starter cultures (1 – 1.5%)

↓

Packaging

↓

Incubation (30° – 37°C/6-8hr)

↓

Dahi

↓

Cooling and storage <5°C

Industrial method of making Dahi:

1. **Selection of raw material:** Production of cultured/fermented milk demands high quality raw materials with respect to physical, chemical and microbial standards.

2. **Filtration/clarification:** Fresh raw milk is heated to 35 to 40°C to aid clarification or filtration process then it is filtered to ensure that, milk is free from extraneous matter.

3. **Standardization (Fat: 0 – 5%, SNF: 11 – 13%):** Fat is standardized based on type of product ranging from fat free to full fat and SNF level is increased by min. 2% than that of milk. It is common to boost the SNF content of the milk to about 12% with the addition of skim milk powder or condensed skim milk. Increased SNF inturn increases the protein, calcium and other nutrients and resulted with improved body and texture, custard like consistency. Higher milk solids prevent wheying off of the product during storage.

MISTI DAHI

Traditional method of *Misti Dahi* preparation: Traditionally, it is prepared by *Halwais* on a small scale to meet the local demand. Whole cow milk or buffalo milk or their combination is added with sugar (upto 15% of the milk) and it is heated continuously in an open pan at a simmering temperature of 68-70°C for 6-7 hours or boiled to reduce the volume to 60-70% of the original. Intense heating imparts cooked flavour and brownish colour to the product. Artificial colour, caramel sugar and/or jaggery are also added based on the consumer preference. The mix is then cooled to about 40°C and inoculated with the previous day's product. It is then filled into earthen pots of consumer size or bulk size vessels and incubated overnight at room temperature. After firm setting of curd, it is shifted to a cooler place or stored under refrigeration temperature. The conditions, under which the product is generally prepared, stored and marketed by the *halwais* are unhygienic. The product is sometimes contaminated with different types of microorganisms including yeasts and moulds which gain entry into the product from utensils and surroundings. Many flavour defects such as fruity, alcohol, highly acidic and flat and textural defects such as gassiness, weak body, wheying off and thick crust on top surface are observed in most of the market samples. These defects may be due to difference in quality of milk, degree of concentration of milk solids, type of culture used, incubation time, processing conditions and temperature of storage. In view of the growing demand, a technology for industrial production of this product has been standardized.

Industrial method of production: In the organized sector a wide range of milk products are used for sourcing milk solids for the production of *misti dahi*. Extreme care is needed in the selection and use of the raw materials and sweeteners. The ingredients should be fresh, good in microbial and sensory quality. Calculated amount of milk and cream is taken into a multipurpose vat. Skim milk powder is added through a venturi assembly to increase the level of total solids and sugar is added at the rate of 9 – 10%. Commercially available Caramel is added normally at the rate of 0.10 to 0.12%. After the mix is prepared, it is heated to 90°C for 10 min in a vat or plate pasteurizer. Then the product is cooled to 40 – 42°C and starter culture containing *Lactococcus lactis* subsp. *lactis* and *Lactococcus* lactis var. *diacetilactis* is added at the rate of 1% to the mix and mixed well. Selection of appropriate type of starter culture is very important and crucial as it affects the flavour, consistency and acidity development in the presence of sugar and caramel. After inoculation the product is packed into sanitized polystyrene cups and sealed airtight. The sealed cups are incubated at 40 – 42°C for about 6 to 8 hours till the acidity develops to about 0.8 % lactic acid.

Once the product develops the desired acidity level of 0.8% lactic acid, it is shifted from incubation room to cold store and maintained at < 5°C. Care should be taken to maintain the temperature of the cold store, so that product doesn't freeze.

Flow diagram for production of *Misti dahi*

Whole milk

↓

Standardization to (3.5% fat 9.0% SNF)

↓

Preheating (65-70°C)

↓

Homogenization* (56 kg/cm^2)

↓

Addition of sugar (@9-10% of milk),
SMP and caramel (0.1 to 0.12%)

↓

Heating to 90°C/10 min

↓

Cooling to 30°C

↓

Inoculation with LF-40 (*Lactococcus lactis* sub sp. *lactis* and *Lactococcus lactis* var. *diacetilactis*) @ 1%

↓

Packaging

↓

Incubation at 30°C for 6-7 hr (acidity of 0.8 to 1%)

↓

Cooling to 4±1°C

↓

Storage 4±1°C

CHAKKA

Traditional method of making chakka: In traditional method, cow or buffalo or mixed milk is boiled thoroughly and cooled to room temperature (30°C). Previous day curd is added to this milk at the rate of 1 to 1.5 %. Milk is left undisturbed overnight at room temperature to set firmly. It is then stirred and hung in a muslin cloth for 10 to 12 hrs to drain off whey. The curd mass obtained after removal of whey is called as chakka.

Flow diagram for traditional method of making chakka

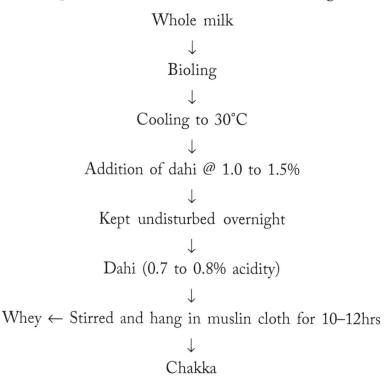

Whole milk

↓

Bioling

↓

Cooling to 30°C

↓

Addition of dahi @ 1.0 to 1.5%

↓

Kept undisturbed overnight

↓

Dahi (0.7 to 0.8% acidity)

↓

Whey ← Stirred and hang in muslin cloth for 10–12hrs

↓

Chakka

Industrial Production of Chakka: Skim milk is normally used in the commercial production of chakka. Low fat in the curd resulted with significant reduction in fat loss through whey, faster moisture expulsion and lower moisture retention in the final product. Fresh, good quality skim milk is received and heated to 90°C for 10 min. High heat treatment kills the competitive microbes and create congenial environment for the growth of culture bacteria. Heat treated milk is cooled to 30°C and inoculated with LF-40 culture containing *Lactococcus lactis* subsp. lactis and *Lactococcus Lactis* var. diacetylactis at the rate of 1.0 – 1.5%. Milk added with culture bacteria is incubated at 30°C for 10-12 hr. After the required acidity of 0.8 to 0.9% LA is reached, the curd is taken into basket centrifuge or quarg separator to remove whey from the curd. Use of quarg seperator for removal of whey has increased the chakka production to 8 tonnes/day. Thus obtained curd mass/chakka is ready for further processing to the production of shrikhand.

Flow diagram for traditional method of making chakka

Skim milk

↓

Heat treatment (90°C/10 min)

↓

Cooling to 30°C

↓

Inoculation with LF-40 culture @ 1.0 to 1.5%

↓

(*Lactococcus lactis* subsp. lactis and *Lactococcus Lactis* var. *diacetylactis*)

↓

Incubation (10–12 hr)

↓

Dahi (0.8 to 0.9% acidity)

↓

Whey Basket centrifuge/quarg separator

↓

Chakka

SHRIKHAND

Traditional method of making shrikhand: Traditionally shrikhand is prepared by boiling cow or buffalo or mixed milk and cooled to room temperature (30°C). Heated and cooled milk is added with previous day *dahi* at the rate of 0.5 to 1 %. Milk is left undisturbed overnight at room temperature to set firmly. It is then stirred and hung in a muslin cloth for 10 to 12 hr to drain off whey. The curd mass obtained after removal of whey is called as chakka. Chakka is then added with calculated amount of sugar, color, flavour and other optional ingredients like fruits, nuts, spices, herbs and served chilled. The chakka obtained from whole milk/ standardized milk has smooth body, whereas the one obtained from skim milk is little rough and dry. When whole milk is used for chakka making, high fat loss occurs in whey thereby affecting the recovery of fat in chakka. Therefore, it is preferred to use skim milk for chakka making and then mixing of cream or unsalted butter to adjust the fat in the finished product. Homogenization of milk leads to slow drainage of whey giving higher moisture content in chakka and a product with very soft consistency (not liked by the consumers). Conventionally made chakka varies from batch to batch with regard to moisture and acidity. Moisture content affects the yield, consistency and composition, whereas acidity affects the taste and quantity of sugar to be added.

Industrial production of shrikhand: With a view to overcome some of the limitations of the traditional method and to partially mechanize the shrikhand production, a semi-mechanized large scale production is employed. Shrikhand is the first traditional milk product for which large scale production technology was adopted. The first modern plant has been established at the Baroda District Cooperative Milk Producers Union Ltd. Baroda Dairy has adopted a process which involves use of basket centrifuge for speedy draining of whey and a planetary mixer for kneading and mixing of ingredients. For industrial production of shrikhand, fresh skim milk is used as a raw material. Use of skim milk has got many advantages: a) Fat losses are eliminated b)

Industrial method of shrikhand manufacture

Skim milk (100 kg) of 9% T.S. 0.13% TA 6.7 pH

↓

Heat treatment 85°C for 30 min

↓

Cooling to 30°C

↓

Inoculation (*L. Lactis* + *L. Lactis* Var. *diacetilactis* is 1.0 to 1.5%)

↓

Incubation for (10 to 12 hrs)

↓

Dahi 0.9% TA & 4.6 pH (0.7 to 0.8% acidity)

↓

Whey ← Basket Centrifuge 990 rpm for 90 min at 30°C

↓

Chakka 20 kg of 25% T.S. (T.S. in chakka is 5 kg)

↓ TA 2.1 to 2.2% pH 4.4 to 4.6

Planetary mixer (35 to 45 rpm for 40 min)

↓

Shrikhand

| Cream 80% fat 2.4 kg, sugar
| @ 80% w/w of chakka,
↓ cardamom @ 1g/kg of chakka

(pH 4.4 to 4.6, T.A. 1.03 to 1.05%, fat 5-6% Protein 6.5 to 7%
Source 40-43%, Ash 0.49% to 0.53, T.S. 57-60%)

↓

Packaging at room temperature and cold storage

Flow diagram for industrial production of Lassi

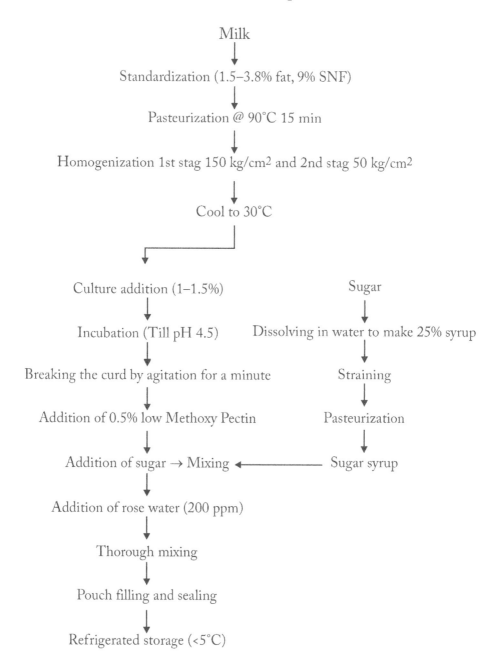

Milk
↓
Standardization (1.5–3.8% fat, 9% SNF)
↓
Pasteurization @ 90°C 15 min
↓
Homogenization 1st stag 150 kg/cm² and 2nd stag 50 kg/cm²
↓
Cool to 30°C

Culture addition (1–1.5%)   Sugar
↓   ↓
Incubation (Till pH 4.5)   Dissolving in water to make 25% syrup
↓   ↓
Breaking the curd by agitation for a minute   Straining
↓   ↓
Addition of 0.5% low Methoxy Pectin   Pasteurization
↓   ↓
Addition of sugar → Mixing ←——————— Sugar syrup
↓
Addition of rose water (200 ppm)
↓
Thorough mixing
↓
Pouch filling and sealing
↓
Refrigerated storage (<5°C)

Flow Diagram for the manufacture of Butter Milk

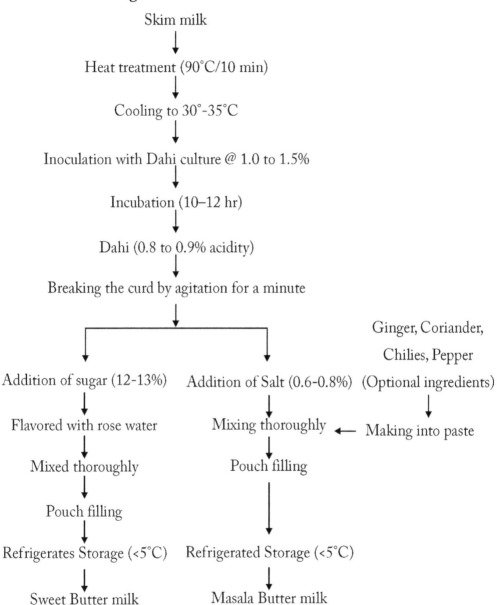

Faster moisture expulsion c) Less moisture retention. Skim milk is heated to 85°C for 30 min, cooled to 30°C and inoculated with LF-40 culture containing *Lactococcus lactis* subsp. *Lactis* and *Lactococcus Lactis* var. *diacetilactis* at the rate of 1.0-1.5%. After the required acidity of 0.8 to 1.0 is reached, the curd is taken into basket centrifuge or quarg separator to remove whey from the curd. The curd mass or chakka is taken into planetary mixer or scraped surface heat exchanger. Sugar at the rate of 80% w/w, calculated amount of plastic cream (80% fat) to give at least 8.5% FDM in the finished product are added and mixed thoroughly. Optinal ingredients like color, flavor, fruits, nuts etc. can also be added at this stage. Then it is packed at room temperature and stored at refrigeration temperature.

LASSI AND CHHACHH/MATTHA (COUNTRY BUTTERMILK)

Method of manufacture: Milk is boiled and then cooled to 30°-35°C. It is added with dahi culture or previous day's dahi at the rate of 1.0 to 1.5%. Milk is allowed to set overnight. Set curd is stirred using a mathani (wooden/steel stirrer with impellers) driven by a small rope in to-and-fro circular motion. During this action, small grains of butter are formed and raised to top of the vessel and it is scooped out from time to time. When all the butter is recovered, the residual watery fluid is referred as butter milk/chhach/mattha/chhas. This can be consumed directly or sweetened or salted and added with spices based on the preference.

Industrial production: Lassi is becoming popular and attracting demand throughout the year. To meet the consumer demand many dairies have started producing lassi on commercial scale. Similar to dahi, fresh, good quality milk is essential for production of good quality lassi. Raw milk is standardized for the fat content ranging between 1.5 – 3.8% and 9% SNF. Standardized milk is heated to 90°C for 15 min, cooled to 60°C and homogenized at 150 kg/cm^2 and 50 kg/cm^2 at 1st and 2nd stage respectively. Milk is cooled to 30 -32°C, inoculated with lactic culture and incubated to attain the pH of 4.5. The curd is broken with the help of a power driven agitator. Sugar syrup (25% syrup) is added to the mix to give 12% sugar concentration in the blend. Low methoxy pectin after making solution in water/syrup can also be added @ 0.5% at this stage as a stabilizer to improve the appearance and mouth feel. The mixture can be flavored with rose water and homogenized to improve body and texture. It is packed and stored at refrigeration temperature.

RAITA, KADHI, DAHIWADA AND RABADI

Method of preparation: *Kadhi* is prepared from *dahi* made from milk standardized to 0.8-1.0 per cent fat. Milk is pasteurized, cooled to 37°C and inoculated with a mixed culture of *L. lactis, S. thermophilus* and *L. cremoris.* This mixture is incubated at 37°C until an acidity of 0.95 per cent (expressed as % lactic acid) is attained. The fermented *dahi* is stirred vigorously in a mixer or using a stirrer. Bengal gram flour at 5 per cent level of *dahi* along with equal quantity of water is added to the stirred *dahi*. Variation in the concentration of Bengal gram flour can change the body and consistency of *kadhi*. After mixing thoroughly, the mixture is cooked at boiling temperature and held at that temperature for 10-15 min. At this stage, appropriate quantity of turmeric powder, spices and salt are added. At the end of boiling, the total solids content will be approximately 14-16 per cent. Research was undertaken to standardize a method for the manufacture of dehydrated *kadhi* or kasha, a ready to mix product to be used for making regular *kadhi* on reconstitution.

RAITA

It is made from *dahi* and served as an additional dish with meals. It is consumed with rice or *roti*. To prepare *raita, dahi* is lightly beaten, spiced and salted to taste. Optional ingredients added to this base include boiled and diced potatoes, raw onion pieces, grated raw cucumber, tomatoes, carrots, pumpkin, ginger, grated coconut and roasted

Flow chart for preparation of dry *kadhi* powder

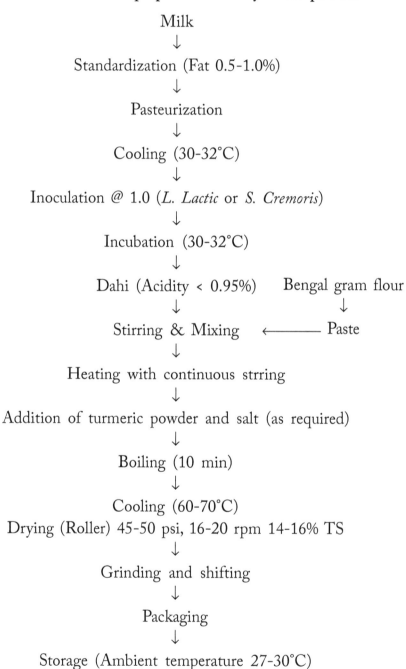

Milk
↓
Standardization (Fat 0.5-1.0%)
↓
Pasteurization
↓
Cooling (30-32°C)
↓
Inoculation @ 1.0 (*L. Lactic* or *S. Cremoris*)
↓
Incubation (30-32°C)
↓
Dahi (Acidity < 0.95%)   Bengal gram flour
↓   ↓
Stirring & Mixing   ←——— Paste
↓
Heating with continuous strring
↓
Addition of turmeric powder and salt (as required)
↓
Boiling (10 min)
↓
Cooling (60-70°C)
Drying (Roller) 45-50 psi, 16-20 rpm 14-16% TS
↓
Grinding and shifting
↓
Packaging
↓
Storage (Ambient temperature 27-30°C)

cumin seeds or fried mustard seeds. Sometimes pieces of fruit like banana and mango may be incorporated after adding sugar and cardamom. *Raita* containing fried *besan* (gram flour) or Black gram *dal* flour granules is particularly popular. Usually, *raita* is consumed in a freshform. At times small quantity of milk is added to *dahi* to develop a soft consistency. It is mixed with salt, black pepper and fried mustard seeds or roasted cumin seeds. The boiled or raw vegetables or *besan /moong* granules are then added and mixed thoroughly. Fruits may be added at this stage. The mixture is garnished with a little red pepper, garam masala and chopped mint / coriander leaves and allowed to stand undisturbed for a few minutes to equilibrate and develop uniform flavour. Depending on regional preferences the quality and optional ingredients added will vary. For example, the ginger curd of Kerala is a thick ginger-based *raita* containing pieces of chopped ginger, green chilles and salt.

DAHIWADA

Dahiwada is eaten as a snack or may accompany a meal as a side dish. To prepare this dish, deep fried black gram *dal* batter patties (*wada*) are dipped in *dahi* and allowed to soak. They are usually garnished with sweet *chutney* prepared from tamarind and jaggery.

Preparation of *Dahiwada:* The ingredients needed are 500 g of black gram *dal*, small amount of red chilli powder, 500 g of *dahi*, frying oil, salt, ginger and other spices. The *dal* is soaked in water overnight. It is then drained and ground to a thick batter, using as little water as possible. It is mixed with spices and shaped into patties of 5-7 cm diameter and 1 cm thickness. Nutritionally rich *wadas* are prepared by placing nut pieces in the centre before frying. The patties are fried in ghee or oil at 150°C until they are cooked properly. They are soaked in salt water for 10 minutes and excessive water is squeezed out. This process assists in the absorption of stirred *dahi,* which is beaten to a thinner consistency after addition of salt and spices. The patties (*wada*) are soaked in the beaten *dahi*. The patties should be covered completely with *dahi* to ensure full absorption of *dahi* liquor by the patties. Before serving, *dahiwada* is garnished with, chilli powder and sometimes with chopped mint leaves. A tamarind sauce is prepared separately by using tamarind pulp and jaggery. *Dahiwadas* are garnished with tamarind sauce before serving.

RABADI

Rabadi is a fermented indigenous food of India especially useful for low and average income rural people who have an easy access to buttermilk. It is popular in North-Western semi-arid regions of India and can be prepared by mixing and fermenting flour of wheat, pearl millet, barley or maize with buttermilk in summer months at room temperature (40-45°C) for 4-6 h. The fermented product is boiled, salted to taste, cooled and consumed. It is a lactic acid fermented food in which lactose undergoes acid fermentation naturally and readily (Gupta, 1989). Cereals/millets have potential to be incorporated in probiotic dairy foods formulation because of their richness in fiber, oligosaccharides, free amino acids and certain minerals which promote the growth of probiotic bacteria. Human-derived strains of *L. reuteri, L. plantarum, L. acidophilus,* and a *L. fermentum* strain isolated from cereals when cultured in malt, barley, and wheat extracts exhibited better cell growth in malt medium than in barley and wheat extracts due to the higher proportion of maltose, sucrose, glucose, and fructose (Charalampopoulos *et al.,* 2002b; Charalampopoulos and Pandiella, 2010).

CHAPTER-7

PROCESSING AND PRESERVATION OF MEAT AND MEAT PRODUCTS

INTRODUCTION

Meat is animal flesh that is eaten as food. Humans have hunted and killed animals for meat since prehistoric times. The advent of civilization allowed the domestication of animals such as chickens, sheep, rabbits, pigs and cattle. This eventually led to their use in meat production on an industrial scale with the aid of slaughterhouses.

Meat is mainly composed of water, protein, and fat. It is edible raw, but is normally eaten after it has been cooked and seasoned or processed in a variety of ways. Unprocessed meat will spoil or rot within hours or days as a result of infection with and decomposition by bacteria and fungi.

Meat is important in economy and culture, even though its mass production and consumption has been determined to pose risks for human health and the environment. Many religions have rules about which meat may or may not be eaten. Vegetarians and vegans may abstain from eating meat because of concerns about the ethics of eating meat, environmental effects of meat production or nutritional effects of consumption.

Processed meats are manufactured from fat and muscle of wholesale cuts, trimmings from carcasses, and some non-muscle cuts such as liver. Processed meats can be divided into fresh processed meat, and cured and smoked processed meat. Ground beef is the most widely recognized example of a fresh processed meat product. Cured and processed meat products make up a large number of the processed meat items sold in the United States and Europe. Curing and salting meat is one of the oldest forms of meat preservation. Most curing procedures use a salt-based brine or pickle for manufacturing the cured meat products. Sodium nitrite is often included in the brine and is responsible for the cured meat color, adds flavor, and can help prevent the development of spores from microorganisms such as *Clostridium botulinum*. Most cured meat products are also heated or cooked in a smokehouse to enhance the flavor and surface color of the meat.

Processed meat is considered to be any meat which has been modified in order either to improve its taste or to extend its shelf life. Methods of meat processing include salting, curing, fermentation, and smoking. Processed meat is usually composed of pork or beef, but also poultry, while it can also contain offal or meat by-products such as blood. Processed meat products include bacon, ham, sausages, salami, corned beef, jerky, hot dog, lunch meat, canned meat and meat-based sauces. Meat processing includes all the processes that change fresh meat with the exception of simple mechanical processes such as cutting, grinding or mixing.

Meat processing began as soon as people realized that cooking and salting prolongs the life of fresh meat. It is not known when this took place; however, the process of salting and sun-drying was recorded in Ancient Egypt, while using ice and snow is credited to early Romans, and canning was developed by Nicolas Appert who in 1810 received a prize for his invention from the French government

SAUSAGES

Ingredients

Meat, salt (2-3%), sugar (for flavor development), spices, nitrates and nitrites, starter culture, encapsulated citric or lactic acid, reducing agents (Ascorbates), antioxidants (BHA or TBHQ), mold inhibitors, soy, wheat, or milk products.

Methods

1. Collect equipment and non-spoilable materials (meat grinder, sausage stuffer, fruit, herbs, spices, salt, natural hog casings, etc.)
2. Get meat. If the meat is freshly killed, you'll only have a few days to get all the processing done before you should have the meat frozen
3. Carve the meat into 1-2 inch chucks suitable for grinding
4. Prepare flavorings
5. Partially freeze meat and flavorings
6. Grind meat and flavorings into sausage
7. Re-partially freeze sausage
8. Stuff sausage into casings
9. Fully freeze sausages

Chicken meat pickle

Recipe: Meat (500 g), fenugreek seeds (2.50 g), cumin (5 g), mustard seeds (5 g), chili powder (15 g), turmeric powder (5 g), salt (15 g), vinegar (100 g), mustard oil (200 g), chicken masala (5 g).

Method. Chicken meat (breast part) was procured from the local market of Kathmandu valley. It was then cleaned with warm water. Then the heat treatment of meat was done by frying and smoking method. The meat was fried in little mustard oil in frying pan at 175±5°C for 10 minutes to golden brown color to prepare fried chicken meat pickle. While wood smoke at 75-80°C, for 30 minutes was done to prepare smoked chicken meat pickle. Thus, cooked meat was cut to desirable size of about 1-2 cm. The spices were mixed and fried in the oil. The spices, salt, and vinegar were mixed with pre-cooked meat followed by light frying. Thus, prepared meat pickle by frying and smoking methods were separately filled in clean dry glass jar up to the neck. This was topped with mustard oil which was heated to 175°C and cooled to room temperature a short while ago.

Meat balls

Ingredients. Meat (minced): 500 g, 2 garlic cloves (crushed), 1 egg, lightly beaten, plain flour for rolling, 3 tsp of sunflower or vegetable oil.

Method

1. Add the minced beef, garlic, parsley, and egg, then mix well. Season well with salt and pepper, and mix again.

2. Dust a large plate or board with a little flour. Scoop out level dessertspoons of the mix, dip them in the flour and roll them into balls. You may find it easier to put a little oil on your hands to help you shape them before you put them in the flour.

3. Heat 3 tbsp of the oil in a large frying pan, and then fry the meatballs in batches, browning them on all sides. Set aside on a clean plate. When you've fried all the meatballs, pour off any excess fat. Rinse and dry the pan.

4. To make the sauce, pour 1 tbsp oil into the pan, heat for 1 min then add 2 crushed garlic cloves and fry for a few seconds. Tip in the tomatoes and break them down as you stir. Cook over a medium-high heat for 5 mins until starting to thicken. Season to taste with salt, pepper and a pinch of sugar.

5. Tip in the meatballs and turn them over in the sauce, ensuring they're all covered. Cover the pan and cook the meatballs on a low heat for about 30 mins. Spoon the sauce over them occasionally, and add a little water if it's becoming too dry. 15 mins before the end of the cooking time, cook the spaghetti.

6. To serve, stir most of the remaining parsley into the sauce, then spoon the meatballs and sauce over the spaghetti. Scatter with the last of the parsley and serve with parmesan on the side.

Dehydrated meat

Meat drying is a complex process with many important steps, starting from the slaughtering of the animal, carcass trimming, the selection of the raw material, proper cutting and pre-treatment of the pieces to be dried and proper arrangement of drying facilities.

Method

1. **Selection of meat for drying.** As a general rule only lean meat is suitable for drying. Visible fatty tissues adhering to muscle tissue have a detrimental effect on the quality of the final product. Under processing and storage conditions for dry meat, rancidity quickly develops, resulting in flavour deterioration. Dry meat is generally manufactured from bovine meat although meat from camels, sheep, goats and venison (antelopes, deer) is also used. The meat best suited for drying is the meat of a medium aged animal, in good condition, but not fat. Carcasses have to be properly cut to obtain meat suitable for drying.

2. **Carcass cutting.** The carcass is first split into two sides along the spinal column and then cut into quarters. Fore and hindquarters are separated after the last rib, thus leaving no ribs in the hindquarter. For suspension the hindquarter is hooked by the Achilles tendon and the forequarter by the last two ribs.

3. **Trimming**. After the quarters are suspended so that they do not touch the floor or anything around them, they are trimmed. Careful trimming is very important for the quality and shelf-life of the final product. The first step is to remove (with a knife) all visible contamination and dirty spots. Washing these areas will spread bacterial contamination to other parts of the meat surface without cleaning the meat. After completing the necessary cleaning of the meat surfaces, knives and hands of personnel must be washed thoroughly. Using a sharpened knife, the covering fat from the external and internal sides of the carcass and the visible connective tissue, such as the big tendons and superficial fasciae, are carefully trimmed off.

4. **Deboning**. It is recommended that this operation should start with the hindquarters and follow with the forequarters. The aim is to remove the bones with the least possible damage to the muscles. Incisions into the muscles are inevitable but only at spots where the bones adhere and have to be cut off. Deboning of the suspended hindquarter should start from the leg and proceed to the rump and muscles along the vertebral column. Deboning of the forequarter must start with cutting and deboning the shoulder separately, followed by cutting off the rib set, together with the intercostal muscles. Deboning of the forequarter is completed by removing the meat from the neck and the breast region of the spinal column.

5. **Cutting into small pieces**. Cut meat into small slices to maximize the area for water removal.

6. **Drying**. Dry in sunlight under clean conditions for 3-4 days.

WAZWAN

Wazwan is a multi-course meal in Kashmiri cuisine, the preparation of which is considered as an art and a point of pride in Kashmiri culture and identity. Almost all the dishes are meat-based using lamb or chicken with few vegetarian dishes. It is popular throughout the

Kashmir. *Wazwan*, the Kashmiri cuisine, is a unique and inseparable component of Kashmiri culture. It comprises from seven to 36 dishes of mutton or beef, chicken, fruits, and vegetables. The important ethnic meat products of wazwan include *kabab, tabak maaz, aab gosh, rogan josh, nate-yakhni, rista,* and *goshtaba.* The ethnic meat products of Kashmiri *wazwan* are popular because of their appealing flavor, texture, and palatability characteristics. However, traditional knowledge of these ethnic meat products in other aspects is not carefully documented. As the demand for ethnic/heritage meat products is ever-growing because of rapid urbanization and industrialization, substantial efforts need to be made to meet such increasing requirements. In addition, because of their popularity, there is a vast potential to introduce them at the national level and promote their export. This review aims to describe processing, quality characteristics, underlying problems, and approaches for the development of some important ethnic meat products of Kashmiri *wazwan.*

IMPORTANT MEAT PRODUCTS OF WAZWAN

KABAB

The origin of *kabab* is credited to the medieval soldiers who used to grill meat on their swords in the open fire. Ethnically the meat used for *kabab* making is lamb, but over the years different types of meats such as beef and buffalo meat have been used as per local and regional taste. *Kababs* are made up of fleshy meat that is minced on stone with a tukni (a wooden hammer)

Ingredients. 150 g mutton (keema minced), 100 g chicken (minced), 2 tsp ginger-garlic paste, 1 tsp onion paste, 1 tsp red chilli powder, 1 tsp coriander powder, 1 tsp cumin powder, ¾ tsp powdered pepper, ½ dry ginger powder (*suanth*), 2 tsp oil, 1 egg yolk, salt to taste, coriander leaves to garnish.

Method

1. In a bowl, mix together the mutton mince and the chicken mince. Mix them with your hands.
2. Add ginger-garlic paste to it along with onion paste, red chilli powder, coriander powder, cumin powder, powdered pepper, dried mango powder, saunth, oil, cashew paste and cream. Mix nicely.
3. Add besan (chickpea flour) and egg yolk to bind. Mix well.
4. Add salt to taste. Mix. Cover and keep in the fridge for 1 hour
5. Skewer the kebabs onto oiled skewers and grill or roast till the outside is nicely browned. Baste frequently with oil while grilling.
6. Once the kebabs are cooked, remove to a serving platter and garnish with onion rings, fresh coriander leaves and lemon wedges.

TABAK MAAZ

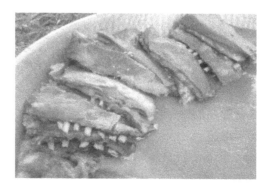

Tabak maaz is a popular product of *wazwan* made from the rib portion of sheep meat. The rib of lamb or mutton is cut into small pieces, moderately steam cooked, and applied with salt and turmeric. The rib bones are then removed, and only the meat chunks are shallow fried in desi ghee over mild heat for a longer period. The finished product becomes crispy in nature and served as dry pieces.

Ingredients. 2 Kg lamb ribs, salt as required, 2 tsp ginger powder, 6 clove, 2 tsp aniseed, 1 cup milk, 500 g ghee, 2 tsp cinnamon powder, 2 tsp turmeric powder, 1 pich asaftoedia

Method

1. Cut the lamb ribs into rectangular pieces. Do not remove the muscles which are covering the ribs. Keep in mind that each rib contains 2-3 pieces of rib bone in the meat.

2. Now, take a vessel, keep it on medium flame and add 1 litre water along with milk in the vessel.

3. Let the mixture of milk and water heat up a little then stir in aniseed, ginger, turmeric, asafoetida, cinnamon powders, cloves and salt to taste.

4. After stirring the mixture properly, add rib pieces and bring the mixture to a boil.

5. Wait till the meat becomes tender and the water is absorbed.

6. Transfer the pieces to a bowl and keep aside.

7. Take a deep bottomed pan, keep it on medium heat and add oil to the pan.

8. Deep fry these pieces in the pan from both the sides and fry till they turn crispy. Serve hot.

AAB GOSHT

Aab gosht is a sacral area of the vertebral column prepared in milk. For the preparation of *aab gosh*, lamb or mutton is boiled in water with salt, ginger, garlic paste, and aniseed powder. The milk is then boiled along with spices such as green cardamom, onions, pepper, and ghee, to which the lamb or mutton is added. The meat along with milk curry is then stirred thoroughly until it boils well.

Ingredients. Milk 250 mL, cardamom crushed 1 tea spoon, cloves-3 numbers, cardamom - 3 numbers, mace- 1 number, salt-to taste, fennel powder- 1 tablespoon, ginger powder- 1 tea spoon, garlic crushed - 4 numbers, mutton - 500 grams, brown onion paste-1/2 cup, pepper powder-1/2 tea spoon, ghee-2 tablespoons, water - as required.

Method

1. Take milk in a pan, add cardomon small powder and cinnamon sticks

2. Boil on low flame till reduce to half

3. Keep aside

4. Now in another pan, take mutton, ginger garlic pastes, onion, green chillies, and spices. Add 1-2 glass of water and cook on low heat till mutton is tender (1-2 hours)

5. When mutton is tender and prepared reduced milk

6. Cook on low heat for further 10-15 minutes on simmering heat

7. At last, add butter and oil and simmer for 5 minutes

8. Delicious *Kashmiri* aab gosht is ready.

GOSHTABA

Goshtaba is a popular meat product of wazwan and is defined as restructured meat product prepared from meat emulsion with added fat (20%), salt (2.5%), cumin (0.1%), and cardamom seeds (0.2%) cooked in the curd. Like *rista*, it is prepared from pounded meat emulsion. The only difference is that *goshtaba* is cooked in gravy called *yakhni* made from curd, water, spices, and condiments. *Rista* and *goshtaba* differ in flavor profile owing to the basic differences in the formulation of gravy. Nowadays, machine mincing is used for the preparation of these products, but the quality of traditionally processed *goshtaba* and *rista* is reported to be superior to that of machine-minced products. To prepare *yakhni*, two parts of fresh curd is homogenized with one part of water with a stirrer, then transferred to a thick-bottomed copper utensil and heated rapidly for 10–15 minutes. During heating, curd is constantly stirred until it reaches the boiling point. Hydrogenated mustard oil is added to the mix, and boiling is continued for 10–15 minutes. Then garlic paste is added followed by a spice mixture. Fried onion paste is added at the end. Boiling is continued until the added oil floats back. At this stage, the remaining water is added, and *yakhni* is cooked further for 10–15 minutes to obtain a desirable consistency. Salt is added toward the end of cooking. The meatballs are then transferred to the boiling *yakhni* and cooked for 30 minutes to get the *goshtaba* in its final form. In the series of different meat products served in *wazwan*, *goshtaba* is the last one to be served. Because of the popularity of *rista* and *goshtaba*, their market potential can be exploited by small-scale industries by packing them in retort cans for long-term storage at ambient temperature or in low-density polyethylene pouches for increased shelf life up to 7 days in refrigerated storage.

REFERENCES

Ait-oubahou, A., 2001. Consultancy Report on Horticulture (Post-harvest and Processing). Winrock International ARMP (TA). Dhaka, Bangladesh.

Alam A 1989." Post Harvest Technology in India" Proceedings of SAARC counterpart Scientists Meeting on Post Harvest Technology ICAR New Delhi, Sep., pp 27-29.

Anon. 2005. "Food Policy- 2005." Indian Food Industry 24 (2) pp 16-20

George J. 2004." WTO and food processing industries in India: Impact and challenges"' Indian Food Industry 23 (1) , pp12 – 20

Hui, Y. H., Ghazala, S., Graham, D. M., Murrell, K. D., Nip, W. (2004). Handbook of vegetable preservation and processing. Marcel, Dekker, Inc. New York.

Ilyas SM.2007. "Processing and value addition through agro processing centers for employment generation." Book Chapter in post harvest management of Horticulture crops. Agrotech publ. Co. Udaipur, pp127-143.

Mehta A, Ranote PS and Bawa As. 2002." Indian Fruit Processing Industry: Quality control Aspects." Indian Food Industry 21 (1), pp37-40

Padda G. S. 2006. "Status and Scope of Food Business in India." Lecture Delivered in winter school on Emerging Trends in Food Technologies and Food Business Development in India in PAU Ludhiana.

Sivasankar, B. (2009). Food Preservation and Processing. PHI Learning Private Limited, New Delhi.

Srivastava, R. P., Kumar, S. (2017). Fruit and Vegetable Preservation. CBS Publishers and Distributers, New Delhi.

UNESCAP, 1997. Post-harvest and Food Processing Technologies. RNAM/ESCAP. Bangkok, Thailand.

For Product Safety Concerns and Information please contact our EU
representative GPSR@taylorandfrancis.com
Taylor & Francis Verlag GmbH, Kaufingerstraße 24, 80331 München, Germany